The Infinite Iteration Principle

Vol. 1: Physics

Brian Capleton PhD

The Infinite Iteration Principle - Vol.1 : Physics

Amarilli Books

ISBN 978-1-917403-17-7

Contents

Preface

The Infinite Iteration Principle – Vast Generative Field (IIP–VGF) framework is a theoretical framework that approaches physical law, mathematical form, and the evolution of complex structure as successive stabilisations of an underlying generative recursion.

While not intended as a replacement for established physics the framework aims to show the possibility of a unified generative architecture within which quantum theory, general relativity, symmetry, decoherence, cosmology, biological evolution, and even intelligence appear as hierarchically nested stabilisations.

At its core the framework studies how stability, redundancy, curvature, and causality arise from the iteration of an abstract operator. The approach draws on insights from fixed-point theory, higher-category theory, quantum information, and tensor recursion, integrating them into a single conceptual model.

The presentation of this booked is aimed at as wide an audience as possible within the demands of the subject matter; we introduce the theoretical basis of the IIP-VGF framework, make some initial demonstrations for its application, and apply the framework to a variety of instances, to show its potential. In the latter part of the book more specific and detailed applications are made.

The IIP-VGF framework presented here has been developed from earlier exploratory work on the infinite iteration principle outside the domain of standard mathematics (*Structures and Morphisms*, 2017 and *Objects and Structures*, 2022). The present IIP-VGF framework is based on standard mathematics and has been made possible through intensive use of artificial intelligence which until now was not available with the necessary power.

Iteration

Iteration is everywhere in the natural and human world. Generally, it simply means repeating a step, a process, or a pattern, often letting the results of one round feed into the next.

Nature herself is iterative: The daily cycle of light and dark, the repeating turns of the seasons, the way rivers carve valleys grain by grain. Biology is built on iteration: Cells divide, DNA copies itself, and small changes accumulate generation after generation.

Human life is equally iterative: We learn through repeated practice, build institutions through repeated actions, and refine technologies by trial, error, and improvement. Across science, mathematics, and society, iteration is the quiet engine that shapes stability out of change, complexity out of simplicity, and progress out of persistence. It underpins both change and stability, often being the means through which change takes place, but also a means through which some kind of thing, a *motif* even, stabilises its existence through the redundancy of iterated instances of itself.

The organisms of a species are an example of such stability through redundancy, but also, iteration is behind evolution itself. It is there in forms such as the cycles of the Earth, the seasons, life, death, and reproduction, and biological iteration in living organisms, from repeat patterns of behaviour to the heart-beat, breathing and brain rhythms. We are used to encountering diverse instances of iteration in science and mathematics. Any cyclical phenomenon is iterative, and so is a simple converging infinite series, in mathematics.

So iteration is ubiquitously encountered as a manifest principle across the sciences and in mathematics. It appears whenever a process, principle, or object is repeated, and when through some functionality it had outputs and inputs, and its outputs are fed back into its inputs, it can revealing stable or emergent patterns, often complex. We see this in recursive definitions and algorithms in mathematics, fractals generated by iterated maps, dynamical cycles in ecology, and feedback loops in engineering control systems.

In physics, iterative approximations underpin perturbation theory, while in biology, cellular replication and generational evolution exemplify iterative processes in nature.

In these contexts, understanding iteration functions as a technique or process applied within existing structures and dynamics. However, in the IIP-VGF framework iteration becomes fundamental.

The IIP–VGF framework is compatible with the core intuition of structural realism — that what science tracks most reliably is structure rather than substance — but it departs from it at a crucial point. Structural realism typically treats structure as ontologically primary: relations, symmetries, or mathematical forms are taken to be what fundamentally exists. By contrast, the IIP–VGF framework treats structure itself as emergent. Structures are not primitive constituents of reality, but stabilised outcomes of a deeper generative process. They are what survives.

Relations do not exist first and then persist; rather, persistence under iteration gives rise to relations that survive and can be treated as structure. In this sense, the framework is neither realist about objects nor realist about structure in the traditional way. It is instead realist about generativity: what is fundamental is the iterative activity from which both objects and structures arise.

Mathematics and structural description then enter retrospectively, as the language of those patterns that survive. Structural realism correctly identifies the durability of structure across theory change, but the IIP–VGF framework explains why such structures are durable in the first place — because as we can now understand them using mathematics, they are the fixed points of iteration under loss of fidelity. But in the IIP-VGF framework nature is not mathematical and then physical; it is generative first, and only later describable mathematically once something has survived.

Quick Shortcut to the Essentials of IIP-VGF

Generative Foundations: The IIP

The basis of the IIP-VGF framework is the Infinite Iteration Principle (IIP): A primitive operation $\mathscr{I}$ iterates on itself without bound.

Stable structures correspond to fixed points and idempotent configurations of $\mathscr{I}$.

No prior metric, topology, or algebraic structure is assumed. These appear as emergent stabilisations under iteration.

This parallels:

- fixed-point attractors in dynamical systems,
- stable categories in higher-category theory,
- pointer states in quantum Darwinism,
- and renormalisation fixed points in QFT.

The IIP is more primitive than these: it presupposes only the existence of iteration.

The VGF

The VGF or Vast Generative Field is the resulting structure:

$$\mathrm{VGF} = \mathrm{Cl}\{\, \mathscr{I}, \mathscr{I} \circ \mathscr{I}, \mathscr{I}^{(n)}, \dots \,\},$$

the closure of the primitive iteration under infinite self-composition.

Within this field:

- Stability corresponds to coherent, recurrence-type structures in the recursion.
- Fidelity corresponds to the environmental redundancy generated by decoherent branches of the iteration.
- Curvature is interpreted as the deformation of the fidelity structure caused by stabilised loops.
- Redundancy generalises "environmental records" in quantum Darwinism.

These map onto physical constructs:

VGF quantity	Physical analogue
Stability	Hilbert-space coherence, quantum states
Fidelity	Spacetime structure, causal cones
Redundancy	Decoherence records / gravitational field
Curvature	Einstein tensor / Ricci curvature
Iteration depth	Cosmic time / renormalisation scale

The QTR: Quadratic Tensor Recursor

The core dynamical equation of the framework is the Quadratic Tensor Recursor (QTR):

$$\mathscr{R}_{\mu\nu} = \alpha\, \mathscr{K}_{\mu\rho}\mathscr{K}^{\rho}{}_{\nu} + \beta\, \mathscr{K}_{\mu\nu} + \gamma\, g_{\mu\nu}$$

where:

$\mathscr{K}_{\mu\nu}$ encodes stability (generalising a quantum density tensor or coherence tensor),

$\mathscr{R}_{\mu\nu}$ encodes fidelity / curvature,

$g_{\mu\nu}$ is the induced metric,

α, β, γ are generative coefficients.

The QTR replaces the Einstein equation:

$$G_{\mu\nu} = \kappa\, T_{\mu\nu},$$

in that it arises before geometry, energy, or stress-energy exist.

Einstein's equation appears as a stabilised limit of the QTR after decoherence cascades.

In the GR limit:

$\mathscr{K}_{\mu\nu} \rightarrow$ effective matter content

$\mathscr{R}_{\mu\nu} \rightarrow$ Einstein curvature

γ term → cosmological constant

QTR → "GR + effective redundancy sector"

Nested Decoherence and the Emergence of Spacetime

A pivotal structural claim in the IIP-VGF framework is:

Spacetime is the first large-scale redundancy stabilisation of the generative recursion.

Before the Planck boundary:

- coherence loops form,
- decoherence propagates at a maximal rate (identified with c),
- redundancy accumulates,
- a fidelity shell emerges.

This redundancy shell is the metric structure, *i.e.*

spacetime = the stabilised redundancy of primordial decoherence.

All later structure — quantum states, classicality, macroscopic objects, thermodynamic irreversibility, biological evolution, cognition — decoheres inside this fidelity shell.

Thus spacetime is the outermost nested decoherence layer.

All physics occurs "inside" the original decoherence that generated curvature.

This reproduces:

- the emergence of a Lorentzian metric from information propagation bounds (QIT approaches to relativity),
- the role of decoherence in defining classical spacetime (Diósi, Penrose, Zurek, Gell-Mann & Hartle),
- causal-set/entropic emergent gravity logic (Jacobson, Padmanabhan).

But in VGF the relationship is strictly generative rather than thermodynamic or entropic.

Expansion as Redundancy Growth

In the VGF interpretation:

$$H(t) = \frac{\dot{a}}{a} = \frac{\dot{R}}{R},$$

with R(t) the radius of the decoherence/fidelity shell.

Hence:

Early Universe: $\dot{R} \approx c$ (coherence propagation dominates)

Late Universe: expansion outruns decoherence; $\gamma g_{\mu\nu}$ term acts as a Λ-like redundancy background.

Thus the ΛCDM expansion law emerges as a redundancy-growth phenomenon:

$$\rho_{\text{eff}}(t) \sim \alpha \rho_K^2 + \beta \rho_K + \gamma.$$

The cosmological constant appears as the asymptotic redundancy term.

The observable relation:

$$H_0 \sim \frac{c}{R_p}$$

arises naturally from the identification of R_p with the particle horizon = decoherence horizon.

Category Theory and c as a Natural Transformation

The framework uses a categorical description:

S : **Gen** → **Stab** extracts stability / coherence data.

F : **Gen** → **Fid** extracts fidelity / curvature data.

Derived functors

$T \circ S$ = time-like scale of coherence,

$R \circ F$ = space-like scale of fidelity.

The speed of light appears as a natural transformation:

$$c : T \circ S \Rightarrow R \circ F$$

ensuring that:

$$R(F(f)) \circ c_X = c_Y \circ T(S(f))$$

for all morphisms f.

This encodes the invariance of causal structure under generative evolution, reproducing the invariant-speed postulate of SR as a naturality condition.

Relation to Mainstream Physics

The framework maps onto established physics in the following ways:

Quantum Theory

- stability = coherence loops in Hilbert-space-like structures,
- decoherence = generation of redundancy,
- pointer states emerge from iterated stability.

General Relativity

- curvature = redundancy tensor $\mathscr{R}_{\mu\nu}$,
- metric = induced fidelity structure,
- GR is a low-coherence, high-redundancy limit of the QTR.

Quantum Gravity

The VGF provides:

- a generative (rather than quantised) origin of spacetime,
- no need for quantisation of the metric (the fidelity shell is not a field to be quantised),
- a natural explanation for the Planck scale as the emergence of the first coherent cycle.

Cosmology

- inflation = early redundancy burst,
- Λ = asymptotic redundancy background,
- horizons = fidelity boundaries,
- Hubble rate = ratio of coherence speed to fidelity radius.

Biological and Cognitive Evolution

Because the generative hierarchy is recursive:

- ecosystems,
- neural systems,
- perceptual classicality,
- memory,
- symbolic intelligence,

all arise as deeper levels of stabilisation inside the primordial redundancy shell.

This is not anthropic, nor does it anthropomorphise physics.

Rather:

The same recursion that produces stability at the quantum and cosmological scale produces higher-order stabilisers at the biological and cognitive scale.

Intelligence is treated as a self-referential stabilisation within the recursion.

Status and Open Problems

The IIP–VGF framework in physics offers potential for further research, with several open directions:

1. In the field equations, rigorous mathematical characterisation of $\mathscr{K}_{\mu\nu}$ establishing precise limits to GR and QFT.
2. Understanding the QTR renormalisation flow.
3. Relating VGF redundancy to entanglement entropy.
4. Deriving symmetry groups as iterated idempotents.
5. Embedding categorical semantics more faithfully.
6. Connecting with quantum thermodynamics.
7. Modelling the transition at the Planck boundary.

Summary

The IIP–VGF framework in physics proposes that:

1. Iteration is the generative primitive.
2. Stabilisations give rise to coherent structures (quantum).
3. Redundancy from decoherence gives rise to curvature (GR).
4. Spacetime is the first large-scale redundancy shell.
5. Expansion is redundancy growth.
6. c is a natural transformation linking time-like and space-like stabilisations.
7. Mathematical structures and physical laws arise from the same generative recursion.
8. (Intelligence is a late-stage stabiliser within the nesting hierarchy).

The framework is a suggested approach to formulating a unified generative foundation beneath existing theories.

The framework is currently in open-form: a formally defined but open structure that describes the constrained manner in which coherence can evolve, without prescribing the substrate or parameters in advance.

Further aspects of the mathematical foundations (including fixed point formalism) can be found in Vol. 6 : *Mathematical Foundations*.

Contextual Background

An Informal Non-Specialist Review

Decoherence

The material world in which we live, considered together with the rest of the universe that astronomy studies, is what physics calls the "classical world". Quantum mechanics, which is often regarded as our most successful scientific theory, shows us that underlying the appearance of this classical world, is what physics calls *quantum coherence*.

Quantum coherence is a very different state of affairs to what we experience and measure as the classical world. When the facts about quantum coherence and its consequences in the form of measurements were first discovered, they were often described as "bizarre" or "weird". For example, it was soon discovered that quantum particles can behave as though they have been in two places at once. Sometimes (in cases of quantum entanglement) they behave as though they have intrinsic connection to each other's states, no matter how far apart in time or space they may be, and without any communication between themselves.

Physicist sometimes talk about "closed" and "open" systems. An "open" quantum system means a quantum system in which there is no differentiation into system and environment, but rather, everything (mathematically speaking, all degrees of freedom that could interact with each other) is considered as a whole.

Limiting an "open" system by partitioning it into "system" and "environment", leads to the current understanding of **decoherence**. What was once considered to be the *mystery* of the "collapse of the wave function" in quantum mechanics, has now been replaced by the understanding of decoherence.

The current theoretical understanding of decoherence doesn't answer everything, but it does show how an otherwise coherent quantum state *decoheres* in interaction with its environment, or under measurement or observation, rather than mysteriously collapsing into classical measurement values. There remains what is called "the measurement problem", which is that decoherence may produce an *multiplet* of possible "pointer states" or measurement values, in which we only have a statistical probability that any one of them will actually be the case in the event of a measurement. Uncertainty still remains, and the root reason for this is still not understood.

Quantum Darwinism

The theory of decoherence has also been extended now into what is called quantum Darwinism. Quantum Darwinism interprets the very robustness of the classical world's objectivity itself, in terms of "pointer states" or the measurable aspects of natural phenomena in the classical world. In quantum Darwinism, what makes the classical world robustly objective, is not that it is not quantum, but that it consists of a vast redundancy of *pointer states,* that comes about through the way natural decoherence copies and broadcasts stable pointer states into the environment.

It's not that quantum coherence or correlations are ever truly lost in nature. Rather, it is that coherence becomes "hidden" and ineffectual, in the "environment". The most important point here is that quantum Darwinism is an explanation in its own terms of why the measurable aspect of the classical world that science studies, has robust objectivity.

So both decoherence theory and quantum Darwinism are all about the distinction between a *system* and its *environment.*

The Partition - the Standard Physics

In standard quantum theory a quantum system can be described mathematically as a density matrix. The density matrix in itself has no split or partition into *system* and *environment.* Mathematically, this distinction is made by partitioning or splitting the density matrix into parts called *system* and *environment*, in a way that is arbitrary.

The consequent mathematics is symmetric, meaning, it doesn't really matter what we call the "system" and what we call the "environment". What matters most for decoherence to effectively take place in nature, is that the "environment" is much larger than the "system".

On making the split or partition in the matrix, the newly designated *environment* part of the original matrix still represents a quantum system with initial coherence, but now it is called the *environment*.

To complete the mathematical modelling of decoherence we apply a mathematical procedure called "tracing out the environment", which is a procedure based on that particular partitioning.

What happens mathematically in the "tracing out", models what happens in nature, which is that a quantum system interacts with its environment, dispersing its coherence into correlations with its environment. This is decoherence. In a *quantum system* there is coherence and entanglement (off-diagonal interference terms in the matrix). But when a quantum system interacts with its environment, the correlations disperse through the environment and become ineffectual.

This remaining entangled state of the *system* and *environment* is not the EPR-style entanglement that has been made popular as an idea in science-fiction. In EPR-style entanglement there is a synchronicity between particles that may be a long way apart in space and time, regardless of how far apart they are. But in physics terms, this happens because there are limited *degrees of freedom* and there are only a couple of particles involved (this kind of entanglement, spin-singlets, *etc.* gives rise to strong, measurable, nonclassical correlations that violate Bell inequalities).

In contrast, the system-environment entanglement that happens in decoherence is of a different nature, because it is usually extremely high-dimensional and distributed. In nature, compared to any *system* state the environment in natural decoherence is *vast*. So there is still quantum correlation or entanglement between the system and the environment, but each *system* state correlates with an enormous cloud of environmental degrees of freedom (in practice, photons, phonons, molecules, *etc*). Any would-be synchronicity, gets completely lost.

The underlying mathematical structure is basically the same, meaning that the *system* and its *environment* are still an inseparably joint state, mathematically, just as the original coherent quantum system was, but now, following the split or partitioning and "tracing out" according to the partition, the details are different. The correlations are now highly proliferate and redundant in the environment, and for all practical purposes impossible to separate out.

So in the mathematics, globally, throughout the whole matrix, coherence is still present, but *locally*, which means from the "point of view" of the designated *system* alone, coherence and entanglement appears to vanish in both system and environment. However, what looks like complete loss of coherence is really the spreading of coherence into a vast array of system–environment correlations, that depend upon the initial split that designates the *system*.

Now the most important thing here, is that this decoherence only happens *relative to a system*. First, there has to be the system-environment split or partitioning.

Quantum Darwinism goes further and explains that the objective measurability of the classical world in which science works, the very fact that it is objective, or the existence of classical scientific "facts" themselves, is synonymous with the vast redundancy of "pointer states" that agree with each other (classical coherence), as the ones that survive and thrive through the process of quantum decoherence - the interaction of a quantum *system* with its *environment*. But this means that as far as our capacity to explain or model theoretically is concerned, the measurable objectivity of our world, the scientific "facts" about it, depend, first of all, on a split or partition.

After the Split

Quantum mechanics and classical science is all about what we can measure. Even measurements made in quantum physics happen in the classical world. Quantum Darwinism says that what is objectively measurable about the classical world, is all about *decoherence*, and a certain stability that arises from the proliferate redundancy of *pointer states* copied into the quantum environment, that quickly makes the classical world objectively measurable.

The morphology of decoherence can be generalised. To begin with, generically, there is a *core.* In *quantum* decoherence this *core* is the indeterministic coherent quantum system. In classical world decoherence it is some deterministic classical system. Then, next, there is a *dissipation* mechanism, or dissipation *channels,* into an "environment", through which the energy or information from the *core* passes, to create decoherence or more robust stability through proliferation and redundancy.

This general morphology happens in both quantum coherence and in classical systems. In the process of *dissipation,* some **fidelity** to the core is lost.

Loss of Fidelity

Quantum coherence itself is fragile, meaning, trying to observe or measure it, disturbs it, or causes its "collapse". Before this, it can evolve in a way that is reversible. After quantum decoherence, the original coherence is effectively lost from the system, but is still "represented" in the environment, where it has become "hidden" and inaccessible.

Quantum coherence is unitary and reversible. It can evolve, but then it can also evolve back to where it was. We cannot, however, get the measurably objective world, which is the result of *decoherence* from the "point of view" of a *system,* to evolve back into coherence.

The general morphology is:

Coherence —(loss of fidelity)→ Decohered State

Quantum System —(loss of fidelity)→ Objectively Measurable World

Reversible Core —(loss of fidelity)→ Irreversibly Stabilised and Robust

The table below captures the universal morphology of change in both quantum and classical domains, structured around two aspects: the reversible core and the dissipative channels.

	Quantum	Classical
Core	Reversible & coherent Unitary evolution preserves full quantum fidelity (phases, superpositions).	Reversible & deterministic Classical drift/ Hamiltonian flow preserves structure without loss.
Dissipation	Decoherence & redundancy Loss of coherence, but redundant proliferation of pointer states across environment fragments.	Dissipation & redundancy Loss of micro-fidelity through diffusion/ noise, but proliferation of stable patterns across redundant channels.

In quantum coherent systems, the key feature is fragile coherence. In the classical world, everything is still basically quantum, but the fragile coherence has been lost in the attainment of robust objectivity. In the IIP-VGF framework this is the loss of **fidelity** to its original state as quantum coherence.

Physics talks about *degrees of freedom*. In both quantum and classical systems the *degrees of freedom* (DOFs) in the core, are channelled into a much greater number of DOFs in the "environment". Degrees of freedom (DOFs) provide a natural measure of dissipation. They quantify the structures and channels through which fidelity is lost and redundancy proliferates.

The Quantum Case

Dissipation occurs when coherence spreads into environmental degrees of freedom.

The decoherence rate Γ is proportional to the density of states (number of DOFs per energy interval):

$$\Gamma \propto g(E)$$

where $g(E)$ is the density of available DOFs at energy E.

The Classical Case

Dissipation arises when energy or information leaks into microscopic DOFs of the medium.

Examples are: Friction spreads kinetic energy into molecular DOFs; diffusion spreads patterns into population channels.
The more DOFs available, the greater the dissipation and redundancy.

General Principle

As dissipation happens, stability arises in the resultant proliferation and redundancy, but fidelity to the original *core* is weakened or lost. What comes to exist through the dissipation becomes a kind of lossy *representation,* in the dissipative "stuff" or "medium" of the original "environment".

The general principle is:

(1) Few DOFs in the dissipation channels → weak dissipation → higher fidelity retained.

(2) Many DOFs in the dissipation channels → strong dissipation → high redundancy and stability, low fidelity.

Thus: dissipation is proportional to the number of available DOFs in the dissipation channels.

DOFs as Proliferation Channels

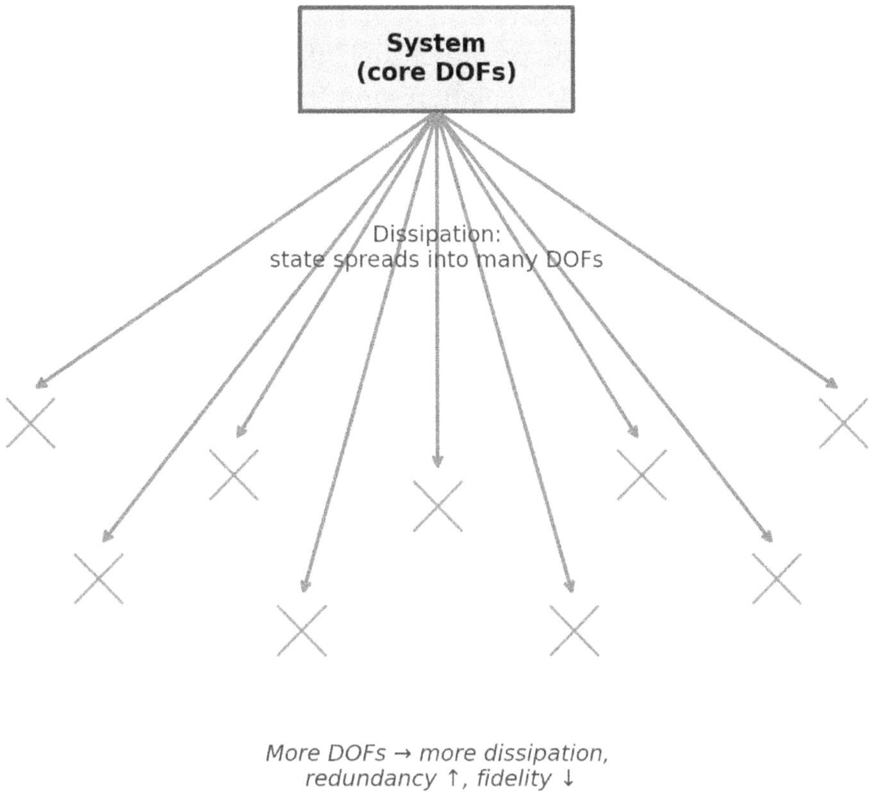

The Quantum Lindblad Equation

The Lindblad master equation in quantum mechanics is one rigorous way of writing a universal morphology of change: a reversible, fidelity-preserving core, plus dissipative channels that spread information redundantly into the environment, thinning fidelity and amplifying stability:

$$\frac{d\rho}{dt} = -\frac{i}{\hbar}[H,\rho] + \sum_k \left(L_k \rho L_k^\dagger - \frac{1}{2}\{L_k^\dagger L_k, \rho\} \right).$$

Here, the reversible *core* is:

$$-\frac{i}{\hbar}[H,\rho]$$

which is unitary, coherence-preserving *iteration*. The dissipative channels $\sum_k(\cdots)$ in the equation are the environment-induced *decoherence* and the formation of redundant records of decohered states in the "environment".

The Classical Analogue

In classical dynamics, the same morphology appears in master-equation or diffusion-like forms:

$$\frac{dp}{dt} = \mathrm{Drift}(p) + \sum_k \mathrm{Dissipation}_k(p)$$

Here, the reversible core is Drift, which is deterministic, fidelity-preserving flow (e.g. Hamiltonian flow, Liouville dynamics, replicator drift).
The dissipative channels ($\sum_k \mathrm{Dissipation}_k$) are noise, diffusion, copying, spreading processes that stabilise outcomes through redundancy, at the cost of fidelity. The quantum analogue is:

$$\sum_k \left(L_k\rho L_k^\dagger - \frac{1}{2}\{L_k^\dagger L_k,\rho\}\right)$$

The General Morphology

The general morphology says:

Evolution = Fidelity-preserving core + Stability-generating dissipation

This captures the Stability–Fidelity Law: Coherence or classical determinism has fidelity to itself at its the core, while environment channels of interaction and dissipation lose fidelity to the core, as they generate stability through redundant proliferation.

The Lindblad equation is the quantum-specialised form of this universal morphology, but the same structure appears in classical systems whenever deterministic drift combines with diffusive or replicative channels.

Schematic View

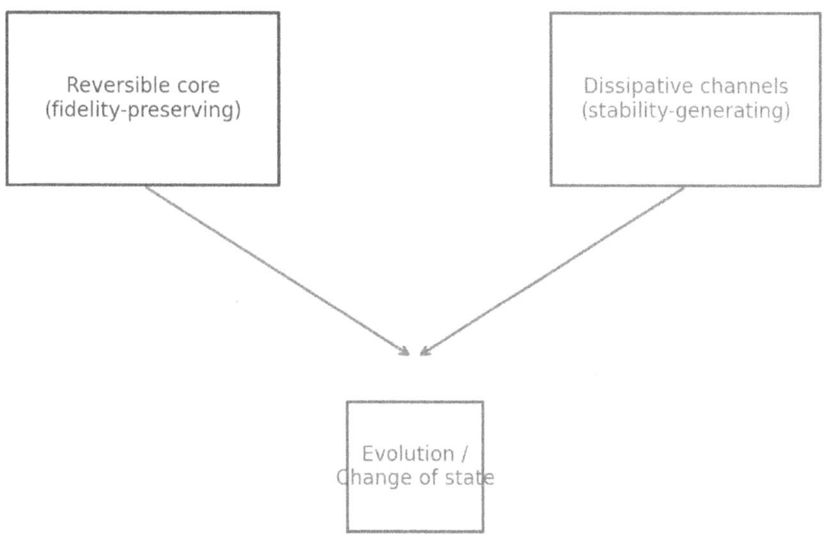

Universal morphology:
Evolution = Core + Dissipation

Quantum vs Classical Morphology

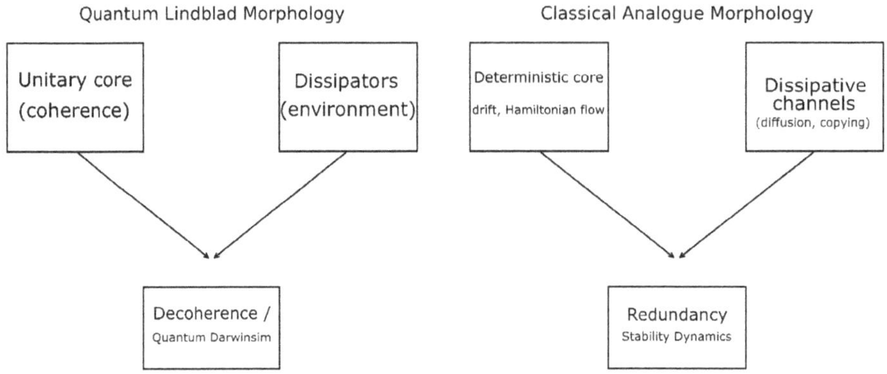

Left: Quantum Lindblad morphology with unitary core (coherence) and dissipators (environment). Right: Classical analogue with deterministic drift core and dissipative channels (diffusion, copying).

Global Coherence in a Closed System

For a closed system (no split or partition), evolution is unitary:

$$|\Psi(t)\rangle = U(t)|\Psi(0)\rangle, \rho(t) = U(t)\rho(0)U\dagger(t)$$

If the initial state is pure, it remains pure for all time:

$$\rho_S E = |\Psi\rangle\langle\Psi|.$$

Thus, globally there is no decoherence, and *coherence* is perfectly preserved. This is what physics calls *unitary evolution*.

Partitioning into Subsystems

Suppose system S interacts with environment E :

$$|\Psi\rangle = \alpha|0\rangle_S|0\rangle_E + \beta|1\rangle_S|1\rangle_E$$

The joint state is pure. But if we only observe the system S, we must trace out E:

$$\rho_S = Tr_E(|\Psi\rangle\langle\Psi|) = |\alpha|^2|0\rangle\langle 0| + |\beta|^2|1\rangle\langle 1|$$

Now the reduced state is mixed — coherence has disappeared locally.

Decoherence is a Limited, Derived Perspective

• Globally: coherence never disappears.
• Locally: decoherence appears when subsystems are defined and parts of the global state are ignored.
• Therefore, decoherence is the natural shadow of entanglement, a "representation" of it, seen from a limited perspective.

Quantum Darwinism Again

Quantum Darwinism *relies* on the subsystem perspective. In Quantum Darwinism:

- Objectivity itself emerges *only* when a system is *first* distinguished from its environment. First there has to be a split or partition.
- Environment fragments then redundantly encode the *pointer states* of the system.
- Without the splitting or partitioning, there is no environment, no redundancy, and no objectivity.

Thus the classical world's objectivity is relational, not absolute. It specifically relates to the partition or split, that relates "system" to separate or distinct "environment".

So decoherence is not a fundamental loss of quantum coherence, and a derived phenomenon, but rather, it is essentially the shadow or "representation" of global unitary entanglement, visible only when we restrict our view to a subsystem.

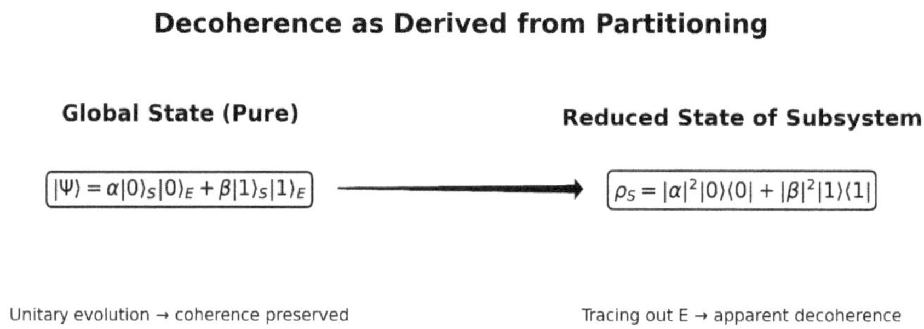

Hilbert Space Partitions

The arena of the mathematics of quantum mechanics is the abstract mathematical space called *Hilbert space*. One of the postulates of quantum mechanics is that Hilbert space *can be partitioned*, in fact, this is the starting point for quantum Darwinism:

$$\mathscr{H}_{\text{tot}} = \mathscr{H}_S \otimes \mathscr{H}_E$$

Quantum Darwinism shows how a quantum *system* interacts with its *environment*, through the interaction Hamiltonian:

$$H_{\text{int}} = \sum_i A_i \otimes B_i,$$

Here, certain system observables A_i commute with this interaction and are left robust. The eigenstates of those observables are the pointer states. In quantum Darwinism *einselection* is then the process by which those states are dynamically chosen as the "stable" basis, because any superpositions of them quickly decohere.

In the Hilbert space partition equation above, the total Hilbert space $\mathscr{H}_{\text{tot}}$ is partitioned into a *system* Hilbert space $\mathscr{H}_S$ and separate *environment* Hilbert space $\mathscr{H}_E$. This is written as though "system" and "environment" were natural, exact divisions. Mathematically, this is an exact factorisation, but it is exact only relative to the chosen split. Hilbert space itself contains no unique, privileged subsystem structure. Any global unitary can scramble one decomposition into another.

No Perfect Split is Inherent in Nature

In field theory, locality partitions leak: the Reeh–Schlieder theorem shows that field states in a region are entangled with the whole. Long-range interactions and cross-couplings prevent perfect separations. In practice, we identify subsystems because certain degrees of freedom interact weakly and can be treated as approximately independent.

Empirical Observations Distilled into Formalism

Quantum mechanics is a set of *postulates*. Our conceptualisation of Hilbert space itself has been distilled from experiments where we already make subsystem complete distinctions such as atom *vs.* apparatus, qubit *vs.* bath, photon *vs.* detector. Thus, when we impose

$$\mathscr{H}_S \otimes \mathscr{H}_E$$

we are not carving along joints inherent in nature, but projecting back into the formalism distinctions that arose from the business of observation in the first place. **Exact splits are the result of our conceptualisation distilled from observations, not inherent joints in nature separate from us**. When we continue to do this in building further theory such as decoherence theory and quantum Darwinism, we are still drawing on the empirical way in which we have deduced the quantum mechanics postulates from observations in the first place.

Now the important point here is that *we* are evolutionary creatures in nature, who have derived this understanding and approach through the situation in which we found ourselves to begin with. Namely, we find ourselves as creatures in a classical environment separate from our organism. In terms of intelligence, we find ourselves as instances of evolutionary intelligence, experiencing and understanding its world as separate from itself. Our current understanding and interpretation of quantum mechanics arises through this scenario.

The Circular Argument Criticism

Decoherence theory is already a successful scientific theory, having been experimentally confirmed to a high degree of precision. Its extension, quantum Darwinism, is currently a well-developed and compelling extension of decoherence theory, with early experimental evidence, but at the time of writing, it is still under active investigation and debate as to whether it is the full story of the emergence of the classical world from quantum coherence.

Both rely on a split or partition. Quantum mechanics itself, the mathematical modelling, in its ability to describe quantum coherence, does not require or overtly suggest a split or partition. Our very conception of distinction between "system" and "environment", comes, in the first instance, from us, and not truly out of the original quantum mathematical model itself.

The physicist and philosopher Ruth Kastner has argued that because decoherence theory *presupposes* a system–environment split, it cannot really "explain" classicality without smuggling in what it wants to explain. To reiterate: One of the ways in which decoherence can be theoretically shown, is by partitioning a density matrix and then "tracing out the environment", which leaves a diagonal matrix in which all the off-diagonal interference terms are suppressed. The diagonal then relates to the probability amplitudes for the possible pointer states that will apply in the event of a measurement.

There is nothing in the density matrix itself that suggests it *should* be partitioned in this way in the first place, and, indeed, the partitioning is arbitrary, and the mathematical process of "tracing out" is symmetrical, meaning that you can just as easily "trace out" the so called "system", as you can the so-called "environment". In nature, as long as the environment is the part sufficiently larger than the system, then decoherence can be shown - the classical world we experience is an instance of it.

The idea that Hilbert space can be partitioned, is a postulate of quantum mechanics, and it works as a method. It is the starting point of quantum Darwinism. But what is really going on here?

Making these partitions is basically just the way in which we humans make further sense of quantum coherence and the presented fact of the classical world, together with the apparent partition between the two. And in that, is the key.

The partition between them only appears in the theory, relative to an already designated *system*. However, we humans *de facto* appear to be stuck on one side of the partition, theory or no theory, and *we* are not included in the theory.

We are creatures who already experience their environment as classical and separate from their organism, and understand the world around them as separate from the intelligence they embody. There is another kind of partition there, in that situation, a **classical partition** between the creature and its environment. There is a partition between the intelligence we are using and what it is understanding, because for everything we understand, we do not yet truly understand the nature of the intelligence itself.

We introduce a **partition** between the representation of a quantum *system* and its *environment*. In doing so, we get the mathematical model to apparently explain *our position*, even though we continue to leave ourselves out of the model. However, the partitions themselves, their deeper role in nature, whether we are talking about the classical partition or the quantum decoherence partition, are **fundamentally unexplained**.

The partition between a creature and its environment is one that, like decoherence itself, and like the classical world, has actual consequences. The universe as a whole, as quantum mechanics tells us, continues to evolve unitarily, without loss of coherence. Even after decoherence, quantum correlations are never truly lost, only the fidelity to their original coherent nature is lost.

The partition between the evolutionary neural intelligence we are using, and what it is understanding, is also not quite what it might appear to be. In fact, there is nothing we are understanding or thinking that is not a construct of our brain function, and *part of* the evolutionary, neural intelligence we are using.

Some people may speculate on how we might include the nature of our own intelligence in the mathematical models of physics one day, in some future scenario. But in fact, essentially, we are *already doing it* by partitioning the models in order to get them to model decoherence, in order to explain the existence of the classical world as we find it through our evolutionary intelligence as evolutionary creatures. And this classical world is of course our world, the world in which organisms such as ours, evolved in the first place.

The IIP-VGF framework addresses these issues by drilling down to the very core of nature's own generativity, and showing that the common generative root in nature, the truly last universal common ancestor of quantum coherence, its decoherence as the classical world, and the coupled evolution of the species and intelligence, is nature's true primitive, understandable in the scientific way, as *iteration.*

The IIP-VGF Framework

Brief Introduction

The principle of continuous, self-recursive iteration, can be postulated as the primitive in mathematics, and can also be postulated as the primitive principle of physics.

This postulated original infinite iteration principle, or IIP, is the abstract principle of recursive self-iteration alone (devoid of any other properties or at attributes).

Examination and analysis then shows that the IIP gives rise to a vast generative field or VGF of relations of relations of iterations, from which springs closures or stabilisations of relations.

The question of relations of relations between objects is the subject of higher category theory, whilst in the IIP-VGF framework objects are treated as emergent from the infinite web of relations generated by the IIP. They emerge as stabilisations, closures, or attractors, in the VGF itself.

General Discussion

The IIP–VGF framework begins from an intentionally austere starting point. Instead of positing particles, fields, laws, or even spacetime, it takes as its sole primitive the idea of iteration — an unbounded, self-applying generative process - nature's generativity.

The infinite iterative process (IIP or Infinite Iteration Principle) is not defined in terms of anything more fundamental. It is simply the postulate that nature's fundamental principle is that something persists in generating. In mathematical terms, it is closer to an operation than an object, a rule that applies to itself, giving rise to an indefinite hierarchy of iterated forms.

From this primitive act of iteration, the VGF (Vast Generative Field) is the closure of everything that can be produced by the generative rule under infinite composition. Formally, if $\mathscr{I}$ denotes the primitive iteration operator, then the VGF is the space of structures invariant under its repeated application. The essential idea is: whatever stabilises under unbounded iteration becomes a structure; whatever does not stabilise is transitory and does not form part of the persistent world.

The Stability-Fidelity Law

Evolution through the Stability-Fidelity law creates the structure of the VGF, in which stable closures of the underlying generative iteration become "fidelity shells". In the framework the earliest stable closure that we can talk about is spacetime. In the IIP-VGF framework, every object and principle treated as objective in science corresponds to a fidelity shell: a stabilised closure that persists by suppressing access to its full generative origin, achieving robustness and public accessibility at the cost of fidelity.

The framework therefore requires a clear sense of what "stabilisation" means. This is where the notion of closure enters. A closure is any structure that becomes idempotent under the action of iteration: applying the generative process again does not change it. Closures are the fixed points of the IIP — they are the first truly durable forms.

Mathematically, a closure E satisfies $\mathscr{I}(E) = E$. Conceptually, a closure is a kind of "environmental shell": a stabilised pattern that constrains further generativity and provides a background for the next layer of structure to arise. In the full VGF, closures form hierarchically: each new closure becomes the environment in which subsequent closures stabilise.

A central organising principle of the framework — one that recurs from physics to biology to cognition — is the Stability–Fidelity Law. This law expresses a universal trade-off: systems that achieve stability do so by sacrificing fine-grained fidelity to the underlying generativity. Stability arises through redundancy, repetition, and coarse-graining, whereas fidelity requires sensitivity to the full complexity of the generative process. The two cannot be maximised simultaneously.

The Stability-Fidelity Law that is central to the IIP-VGF framework can be expressed in a completely generic way as a monotone trade-off under a coarse-graining map.

Let $\mathscr{G}_\lambda$ be a coarse-graining (or "environmental averaging") map at scale λ (classically: a Markov kernel / lumping operator; quantumly: a CPTP channel). Measure "fidelity to the fine/coherent description" by how much distinguishability survives, and "stability" by how contractive the map is.

A very generic, succinct form is:

$$\kappa(\mathscr{G}_\lambda) := \sup_{\rho \neq \sigma} \frac{D\big(\mathscr{G}_\lambda(\rho), \mathscr{G}_\lambda(\sigma)\big)}{D(\rho, \sigma)} \leq 1 \qquad \text{(data processing)}$$

where D is any divergence/metric that obeys data processing (*e.g.* relative entropy, trace distance, *etc.*).

Then define (one convenient choice)

$$\textbf{Fidelity}(\lambda) := \kappa(\mathscr{G}_\lambda), \qquad \textbf{Stability}(\lambda) := 1 - \kappa(\mathscr{G}_\lambda).$$

The Stability–Fidelity Law in coarse/fine-graining form becomes:

$$\lambda \uparrow \quad \Rightarrow \quad \kappa(\mathscr{G}_\lambda) \downarrow \quad \text{so} \quad \textbf{Stability}(\lambda) \uparrow, \textbf{Fidelity}(\lambda) \downarrow.$$

Interpretation:

Stronger coarse-graining (more redundancy/averaging) makes macrostates less sensitive to micro-perturbations (more stable) precisely because it suppresses fine-grained distinctions (lower fidelity to the underlying coherence).

We could also write this as:

$$I_{\text{ret}}(\lambda) := I(\text{micro}; \text{macro}_\lambda) \downarrow \iff S_{\text{rob}}(\lambda) \uparrow,$$

i.e. retained micro-information decreases as robustness increases under coarse-graining.

Quantum Channel Application

In quantum physics, the Stability–Fidelity Law follows from CPTP contractivity: any channel that stabilises states by coarse-graining must reduce trace-distance distinguishability, gaining robustness precisely by losing fidelity to underlying quantum coherence.

The Stability–Fidelity Law in Quantum Channel Form)

Let

$$\Phi : \mathscr{B}(\mathscr{H}) \to \mathscr{B}(\mathscr{H})$$

be a quantum channel (completely positive, trace-preserving), representing environmental coupling, decoherence, or coarse-graining at some scale.

Define quantum distinguishability by the trace distance

$$D(\rho, \sigma) := \frac{1}{2}\|\rho - \sigma\|_1 \,.$$

By the data-processing (contractivity) theorem, for all density operators ρ, σ,

$$D(\Phi(\rho), \Phi(\sigma)) \;\leq\; D(\rho, \sigma).$$

Define the fidelity-retention coefficient

$$\kappa(\Phi) \;:=\; \sup_{\rho \neq \sigma} \frac{D(\Phi(\rho), \Phi(\sigma))}{D(\rho, \sigma)} \;\in\; [0,1]\,.$$

Then the Stability–Fidelity relation may be written as

$$\textbf{Fidelity}(\Phi) := \kappa(\Phi), \qquad \textbf{Stability}(\Phi) := 1 - \kappa(\Phi)\,.$$

Stronger decoherence or coarse-graining corresponds to channels with smaller $\kappa(\Phi)$: they suppress fine-grained quantum distinctions (lower fidelity) while producing outputs that are increasingly insensitive to microscopic variation (higher stability).

Canonical decoherence example (pointer-basis attraction)

Let Δ be the dephasing channel in a pointer basis

$$\{|i\rangle\} : \Delta(\rho) := \sum_i |i\rangle\langle i| \, \rho \, |i\rangle\langle i| \, .$$

- Off-diagonal coherence is eliminated.
- Diagonal states form a stable fixed-point algebra.
- $\kappa(\Delta) < 1$, expressing loss of fidelity to coherent phase relations.

A continuous family,

$$\Phi_p(\rho) = (1 - p)\rho + p\,\Delta(\rho), \qquad p \in [0,1],$$

makes the trade-off explicit:

$$p \uparrow \Rightarrow \kappa(\Phi_p) \downarrow \Rightarrow \textbf{Stability} \uparrow , \textbf{Fidelity} \downarrow .$$

Decoherence

Closely connected to this is the principle of evolution through decoherence. Decoherence, in this framework, is not limited to quantum physics; it is a general mechanism by which the fine-grained products of generativity collapse into coarse-grained, redundant, and therefore stable forms.

Decoherence is the process that transforms unbounded possibilities into persistent, classical-like structures. At the earliest levels of the universe, decoherence of quantum gravitational superpositions produces the first major closure: classical spacetime, complete with causal order. At biological scales, decoherence produces stable organisms from a flux of molecular processes. In cognition, decoherence results in the stabilised schemas, concepts, and memories that constitute mental life. In every case, evolution proceeds by decohering unstable generative possibilities into stable closures capable of further refinement.

Together, these principles — iteration as primitive, closure as stabilisation, the Stability–Fidelity Law, and evolution through decoherence — provide a coherent mathematical and conceptual foundation for understanding how structure arises from generativity. They articulate how the universe, life, and intelligence can all be viewed as nested layers of stabilised iteration, each emerging from — and constrained by — the one beneath.

The IIP–VGF framework thus offers a unified way of understanding emergence across physics, biology, cognition, and mathematics as the natural consequence of the same fundamental generative process. Taken together, **stability** and **fidelity** describe the twin conditions under which order persists in an evolving universe.

Every physical system - whether a quantum state, a planetary orbit, or a field configuration - engages in a relationship between its **stability** or resistance to disturbance or even destruction, and the **fidelity** of its own internal correlations.

The universe itself can be read as the grand negotiation between these two demands: from the coherence of quantum amplitudes to the geometric fidelity of spacetime, all persistence depends on how stability and fidelity weave together.

These terms will acquire more formal meaning for physics as we proceed: **Stability** as the mathematics of local coherence, **Fidelity** as the geometry of global consistency.

The Stability–Fidelity Law expresses a universal tension in all evolutionary process: any process or system that achieves greater stability through redundancy, averaging, or projection, necessarily loses fidelity to an originating state from whence it came. Conversely, maximal fidelity implies fragility, instability, and rapid decoherence. In physics, this principle is evident from quantum to cosmological scales.

A quantum superposition is a high-fidelity carrier of information about coherent phase relations, but is unstable in the presence of environmental interaction. Decoherence - through which classical objectivity arises - stabilises the system by distributing its information redundantly into the environment, at the cost of erasing the original coherence. Fidelity to the original coherence or correlations, has been lost.

Thermodynamic irreversibility, renormalisation in field theory, and attractor formation in nonlinear dynamics all instantiate the same trade-off: fine-scale information is sacrificed to yield coarse-grained invariance and macroscopic predictability. The observer's stable "world" is, in this sense, a product of controlled fidelity loss.

In mathematics, the law appears wherever infinite or unstable structures are regularised. The discretisation of a continuum, the truncation of a divergent series, or the projection of a high-dimensional operator onto a lower subspace each stabilise a representation while shedding fidelity to the original ideal form.

Functional analysis formalises this through the passage from exact functionals to bounded operators; topology and category theory do so through quotienting and equivalence classes that preserve structure-invariant relationships but discard unique instances. In numerical computation, rounding and approximation schemes stabilise iterative procedures by introducing controlled loss of precision — a mathematical decoherence in algorithmic form.

Viewed together, these examples reveal a unifying principle:

> New representations or conditions or states that become **stabilisations** in evolutionary processes, whether in quantum or classical evolution, emerge from past conditions but require selective "forgetting" (of former correlations).

The IIP-VGF framework embraces this principle through the totally generalised principle of **infinite iteration**. How this becomes possible and the full meaning of it will become clear in due course.

Thus, what in physics appears as the decoherence from fragile quantum coherence to stable classical objectivity, and what in mathematics appears as the abstraction from instance to invariant, are seen in the IIP-VGF framework as parallel phases of the same iterative process of stabilisation through decoherence.

So we shall see in due course that the Stability–Fidelity Law can be seen as having its origins in the infinite iteration principle, or IIP. The

IIP-VGF framework begins by postulating the IIP as the primitive principle of nature, and then this naturally follows through to an understanding of evolutionary **decoherence** in nature, as quantum evolution, classical evolution, the evolution of the species, and of intelligence. Thus, decoherence is an expression of the Stability–Fidelity Law operating in all kinds of natural evolution.

In the IIP-VGF framework, Fidelity to a given coherence system has to be given up in order to attain Stability in a new coherence scheme. This is also the Stability-Fidelity Law in action, in what constitutes both an evolution and a decoherence.

We see this in paradigm shifts in science, such as from Newtonian mechanics to Relativity, or from classical physics to quantum mechanics. The paradigm shift is a change in the stabilisation scheme through which science balances robustness against fidelity. The paradigm shift or evolution in theory is triggered when the existing scheme can no longer preserve coherent structure across expanding observational domains. In each case, the new paradigm does not refute the old. It extends the region in which stability and fidelity can be jointly satisfied, revealing the earlier paradigm as a restricted stabilisation valid only within certain limits.

The sense of nesting of theories as understanding evolves, with each shell pertaining to a sphere in which knowledge and ignorance is balanced, reflects the nested nature of the VGF.

The VGF

From the principle of infinite iteration, or IIP, arises the infinite web of relations between relations between iterations - a generative field (vast generative field or VGF). Although the VGF can be considered as a graph, the approach that must be taken is necessarily somewhat different to the one we usually take in graph theory, in that the graph must be given *edge primacy*.

This is dealt with in more detail later in the chapter on this subject. It is in the *relations of relations* (an area also dealt with by higher category theory but in a different way, as we shall discuss in due course) and the fact that it is from these relations of relations arise from the axiomatic principle of **iteration**, that the richness arises.

The relations or edges in the VGF are not merely between static relata as nodes, but between iterations themselves — and this is precisely what distinguishes the VGF from any classical or ordinary categorical ontology. This is why the VGF, as we shall be seeing, will point to subsequent **structures**, including all kinds of numbers, Hilbert space, decoherence, and neural loops.

At first sight, or under a superficial glance, the Vast Generative Field (VGF) seems immune to spectral analysis. If it is a web of infinite relations between infinite iterations, it appears to be a fully connected graph — a complete network in which every node touches every other.

In ordinary graph theory, such a graph has trivial spectra: all nodes are equivalent, and the Laplacian's eigenvalues collapse to a single value.

How, then, can a concept like saturation or Planck frequency make sense within a field that is already maximally connected?

The answer lies in the distinction between potential and effective connectivity, and in the idea that saturation in the VGF is informational, not structural. A more detailed account of this is given in the chapter ***Decoherence and Stabilisation***.

Potential *vs*. Effective Connectivity

In the VGF's p**otential connectivity** every relation-point (node) could, in principle, connect with every other. This defines the infinite generative capacity of the VGF — its raw **coherence field**.

However, its **effective connectivity** is only a subset of those potential relations that **remain stable at a given level of redundancy**, under **continual iteration**. These constitute a **coherent network** that exists at that particular stage of stabilisation.

The VGF from its inception immediately becomes not a fixed complete graph, but rather, a recursive, nested hierarchy of **effective graphs**, each representing one layer of coherence within the infinite potential background.

This principle (the vast generative field or VGF), when we understand the arising and nature of closures or stabilities within it, provides us with a demonstration and understanding of the **principle**

of decoherence - *the loss of coherence to a new coherent stability* in a most generalised form.

The IIP is the root of this principle of decoherence, in which coherent stabilities (loops or closures) form and decohere, providing the basis for new coherences, which may then decohere, and so on. As decoherence takes place, fidelity to the original coherence is lost, but a new kind of fidelity is stabilised. It is an ongoing process.

With the IIP as the primitive, the very principle of decoherence can be seen as the fundamental dynamic, with **quantum decoherence** just one of its stages. The underlying law of the VGF is the Stability-Fidelity Law, in which decoherence into Stability takes place at the cost of Fidelity.

In the IIP–VGF framework, an **attractor** denotes a dynamically stabilised pattern of iteration: a region of generative activity toward which a wide range of microscopic trajectories converge, and within which further iteration reproduces the same macroscopic structure.

Attractors are not static entities but self-maintaining closures, sustained by redundancy and invariant under continued interaction. Within this framework, decoherence may be understood as the bifurcation of a previously unified attractor into multiple stable attractors.

Each resulting attractor constitutes a self-consistent iterative process, dynamically autonomous and effectively orthogonal to the others, such that interference between them becomes impossible under forward iteration. This orthogonality, however, does not represent a fundamental separation. The attractors remain tenuously connected through their common generative origin in the Vast Generative Field itself.

Decoherence thus marks not an ontological branching of nature, system or environment, but a stabilisation-driven partitioning of generative dynamics into mutually non-interfering closures.

What Saturation Means in the Context of the VGF

In a finite graph, saturation means adding edges no longer increases the number of independent cycles.

In the VGF, saturation occurs when adding more *coherence relations* yields no new **distinguishable loops**.

This happens when coherence becomes so dense that further refinement only reproduces existing patterns — when fidelity (to iteration) outruns distinction.

At that threshold, the spectrum of effective connectivity ceases to expand.

The density of independent eigenmodes of the Laplacian operator $L = D - A_{\text{eff}}$ reaches a fixed asymptotic form.

This point of informational closure corresponds morphologically to the Planck frequency: the highest rate at which the field can generate new distinct iterations.

Morphological Table

Concept	Finite Graph	VGF Interpretation
Node	Discrete vertex	Relation-point (potential distinction)
Edge	Binary connection	Coherence relation between distinctions
Loop	Cycle of edges	Self-stabilising iteration
Saturation	All edges added	All distinguishable relations realised
Laplacian eigenvalue	Mode frequency	Iterative coherence frequency
Spectral cutoff	Smallest cycle	Limit of distinction (Planck regime)

Review

- The VGF is infinitely connectable but relationally saturated — further coherence as a network or graph adds no new distinctions. Its spectral structure reflects not how many edges exist, but how many differences survive infinite iteration.
- Spectral saturation marks the limit where further connection yields no new distinction — the closure of effective diversity.
- The Planck frequency is taken to be the upper bound of this informational spectrum: the point where the generative field,

though infinitely relational, becomes self-similar and time itself is a motif.

Stabilisation of Attractors

We can initially model the emergence of stability from repeated self-interaction in the Vast Generative Field (VGF) as:

$$A_{n+1} = (1 - \alpha)A_n + \alpha\, \phi_{\tau,\kappa}\big(\mathrm{Norm}(A_n^2)\big)$$

A_n	Current state (a vector, operator, or local patch of the field)	The *current configuration* or *relational network* inside the VGF at iteration n.
A_{n+1}	Next state after one iteration	The *updated configuration*, after interaction and normalisation.
A_n^2	Self-interaction term	Represents **feedback or recursion** — the system acts upon itself (like $z \mapsto z^2 + c$ in the Mandelbrot iteration).
$\mathrm{Norm}(A_n^2)$	Normalisation operator	Ensures the field remains bounded, analogous to renormalisation or normalisation of amplitude in Hilbert space.
$\phi_{\tau,\kappa}(\,\cdot\,)$	Transformation function depending on parameters τ, κ	Encodes the *form* of internal transformation (e.g., phase rotation, kernel mapping, or nonlinear filtering). Think of it as the "local physics" or morphogenetic rule.
α	Mixing coefficient ($0 \le \alpha \le 1$)	Determines **how much of the new self-interaction replaces** the old configuration. When α is small, the system is inertial (memory-preserving). When α is large, it updates rapidly (more decoherence).

This essentially represents:

- the previous state $A_n \rightarrow$ continuity and memory (stability),

- and the self-iterated transform $\phi(\mathrm{Norm}(A_n^2)) \to$ change and self-generation (fidelity).

Thus, it expresses precisely the Stability–Fidelity trade-off at the heart of the VGF:

$(1-\alpha)A_n =$ stability term (redundant continuation),

$\alpha\, \phi(\mathrm{Norm}(A_n^2)) =$ fidelity term (new fine-structure iteration).

Morphology

It can also be thought of as a recursive averaging map:

- If $\alpha = 1$: full fidelity iteration — pure IIP mode (chaotic, unstable).
- If $\alpha = 0$: full stability — frozen, no change.
- Intermediate values: self-organising behaviour — the zone where attractors form, i.e. where coherent classical structures appear.

Attractor formation happens when this iterative rule converges or oscillates stably, *i.e.*

$$A_{n+1} \approx A_n, \quad \text{or} \quad A_{n+k} \approx A_n,$$

meaning the dynamics settle into a limit cycle or fixed point.

That's the stabilisation event corresponding to the emergence of an object, pointer state, or classical measurable.

Connection to Other Frameworks

Framework	Equivalent structure	Analogy
Quantum HEOM	Recursion of auxiliary density operators $\rho^{(n)}$	Nested memory and feedback terms encode coherence loss.
Mandelbrot iteration	$z_{n+1} = z_n^2 + c$	Same quadratic self-map structure; boundedness criterion defines attractor vs divergence.
Neural dynamics	$x_{t+1} = (1-\alpha)x_t + \alpha f(Wx_t)$	Standard recurrent network update rule — balance of memory and transformation.

VGF IIP	$A_{n+1} = (1 - \alpha)A_n + \alpha\, \phi(\mathrm{Norm}(A_n^2))$	Universal recursive rule describing iteration of iteration — stabilisation through redundancy.

Mapping Analysis

Each iteration of the field takes the old configuration, squares it (self-interaction), normalises it to prevent runaway amplification, and blends it back into the old form.

Over infinite repetition, this produces attractors — stable forms of redundancy — which are the "objects" of the VGF.

Relation Between Objects and Iteration

In conventional mathematics, a relation $R(a, b)$ links two objects a, b taken as given. In the VGF, by contrast, **there are no primitive objects** — only iterations acting upon iterations. **Objects are emergent.**

So the fundamental relation looks like:

$$R(\textbf{iteration}_i, \textbf{iteration}_j)$$

and not $R(x, y)$ between "things".

That means each edge is a connection between generative acts, not between finished states. Edges are mappings of iteration onto iteration — one act of differentiation referring to another.

What the VGF encodes, therefore, is the recursion of recursion itself. The VGF does not merely host recursive dynamics — it is the recursion of recursion itself.

Objects from IIPs

The principle can be applied purely mathematically, or to other fields of science. Each stabilised form, whether mathematical, physical, or cognitive, is an encoding of an iterative dynamic within the context of the VGF. Every object of experience, every measurable, and every act of distinction represents a finite stabilisation of an infinite iterative process.

Each act of **stabilisation** within the VGF constitutes a **local completion** — a finite contraction of infinite recursive potential into a distinguishable **relation-structure**.

Formal Representation

Formally, if iteration is represented as the operator ⊙ acting on itself, then the VGF represents the closure of this process under infinite composition:

$$\odot \ \rightarrow \odot(\odot) \rightarrow \odot(\odot(\odot)) \rightarrow \ \ldots \ \rightarrow \ lim \rightarrow \infty \ \ \odot(\odot)$$

Each stabilised form within this infinite composition corresponds to an idempotent or fixed-point structure, denoted generically as

$$E_i = \text{▯}(E_i)$$

Interpretation

The equation above formalises stability within the infinite iterative field: It defines each E_i as an idempotent or fixed-point structure — an entity that reproduces itself under the universal act of iteration.

In physics, such fixed points correspond to pointer states or stable equilibria emerging through decoherence, where repeated environmental interaction leaves the observable unchanged.

In biological evolution, they represent species or morphological attractors — configurations that persist through adaptive cycles by maintaining functional coherence under environmental iteration.

In cognition, they appear as conceptual invariants or memory schemas — stable representations that remain recognisable despite recursive modification through perception and recall.

Across all domains, $E_i = \text{▯}(E_i)$ expresses the defining feature of stability under transformation: the system has become its own iteration, conserving relational coherence within the vast generative field even as fidelity to its generative origin has been thinned.

These idempotent stabilisations together constitute the structured fabric of the VGF, through which the infinite iteration process becomes manifest as the coherent order in and structure of the discriminatory observable world.

Such idempotents form the **attractors** of the VGF — self-consistent loops of recursion.

The quantum domain can thus be considered as arising in this way. When quantum *decoherence* is introduced, these attractors become redundantly stabilised, forming the observable objects of the classical domain.

Thus, every measurable entity — whether physical (a particle), mathematical (a number), or even cognitive (a concept) — is a stabilised encoding of the recursion of recursion.

This gives rise to the **Stability–Fidelity Law**: fidelity to the generative recursion is lost as stability (redundancy) is gained.

Decoherence, in this context, is the contraction of recursive self-reference into finite, redundant representation. Quantum decoherence is itself then a particular representation of the more general decoherence principle in the VGF, of stabilisation and redundancy at the expense of fidelity to an underlying iterative dynamic.

The Canonical Form of the VGF Dynamics

The dynamics of the Vast Generative Field (VGF) are captured in their most compact and universal expression by the quadratic tensor recursion map

$$g(\mathscr{K}) = \alpha\, \mathscr{K}^2 + \beta\, \mathscr{K} + \gamma\, I,$$

where $\mathscr{K}$ denotes the generative operator acting on itself, and α, β, γ are scalar or tensorial coefficients representing the coupling, persistence, and invariance of the recursion respectively.

This form is canonical because it expresses, in the most economical algebraic language, the three essential processes that any self-iterating field must undergo: nonlinear self-coupling, linear continuity, and invariant grounding.

- Nonlinear self-coupling ($\alpha\mathscr{K}^2$) — Every generative system is defined by feedback: it acts on the results of its own prior action. The quadratic term encodes this reflexivity. It represents the mutual influence of the field's internal and external aspects—the "system" and its "environment"—and is therefore the algebraic expression of iteration itself. Without a quadratic self-coupling, the field would evolve linearly and could not produce novelty, differentiation, or stabilised structure.
- Linear persistence ($\beta\mathscr{K}$) — A self-generating field must also retain continuity across iterations. The linear term represents this persistence: it carries forward the current stabilisation, preserving coherence between successive states. It embodies the memory or inertia of the field, ensuring that change is not chaotic but accumulative.
- Invariant baseline (γI) — Finally, every iteration presupposes a background of identity, the logical or physical "space" within which transformation occurs. The identity term supplies this invariant reference, the conserved quantity or normalisation against which all evolution is measured.

Taken together, these three components form the minimal closed algebra capable of describing the evolution of a self-referential generative process.

The map is quadratic, rather than of higher order, because self-interaction (action of the operator on itself) is the simplest nonlinearity sufficient to produce differentiation and feedback; and it is tensorial, because the field's evolution involves coupling between multiple interdependent channels — stability and fidelity, internal and external, real and imaginary components.

In this sense, the quadratic tensor recursion is the parsimonious canonical form of VGF dynamics. It stands to generative systems as the Schrödinger equation stands to quantum mechanics or the logistic map to population dynamics: the simplest expression that preserves reflexivity, continuity, and invariance while remaining closed under its own action.

All higher-order formulations or domain-specific equations — whether in physics, biology, or cognition — can be viewed as elaborations or particular instantiations of this same universal recursive grammar.

Thus,

$$g(\mathscr{K}) = \alpha\mathscr{K}^2 + \beta\mathscr{K} + \gamma I$$

is not an arbitrary choice, but the natural and minimal mathematical structure for describing a field that iterates upon itself. It is the algebraic skeleton of the VGF: a single form capable of generating both stability and transformation, coherence and decoherence, the persistent self-organisation of the field through its own infinite iteration.

Technical Note: Closure and Idempotence

From a formal standpoint, the quadratic tensor recursion

$$g(\mathscr{K}) = \alpha\mathscr{K}^2 + \beta\mathscr{K} + \gamma I$$

arises as the minimal polynomial closure of a self-referential operator.

Any self-acting process $\mathscr{K} \mapsto f(\mathscr{K})$ that remains stable under its own iteration:

$$\mathscr{K}_{n+1} = f(\mathscr{K}_n)$$

must admit fixed-point or idempotent solutions E satisfying:

$$E = f(E) = E^2.$$

The smallest algebra of endomorphisms closed under both linear combination and self-composition therefore contains precisely the powers $\{I, \mathscr{K}, \mathscr{K}^2\}$.

Higher-order terms $\mathscr{K}^n$ for $n > 2$ add no new structural degrees of freedom, since any iterative hierarchy of self-maps reduces, under idempotent completion (Karoubi envelope), to compositions of these three.

Hence a general second-degree polynomial is the canonical closure of reflexive iteration.

In categorical language, this corresponds to the Karoubi-complete envelope of a self-functor: the construction that adds all stable idempotents and closes the category under its own composition.

The coefficients α, β, γ then parameterise how much of the next morphism is drawn from the quadratic self-interaction, from linear persistence, or from the identity inclusion.

Because this polynomial is both algebraically closed and dynamically sufficient to generate fixed points, attractors, and bifurcations, it serves as the universal normal form for VGF evolution.

Review

The proposition of the VGF as the encoding of recursion of recursion means that every form in the universe — every pattern, equation, or thought — is a trace of an infinite self-iterating act. The world we observe is thus composed of "stabilised echoes" of that iterative principle.

The stable VGF loops that we call "objects" or "laws of nature" are the condensed results of endless iteration, held in balance by redundancy and the place of any thing in the networks of which it is a part.

In essence every object can be considered as a "memory" of **generative activity**, in one form or another. Every object appears in a context of other objects, at its particular level of stabilisation from the infinite iteration principle and its VGF.

So nothing in the universe is genuinely static or isolated; rather, everything is a slowed, stabilised contraction of infinite recursion.

Objects and even mathematical relations and concepts, persist because their internal relations reinforce one another just as a standing wave persists through repeat-pattern of resonance, or a stable idea persists through repeated recall.

The IIP-VGF approach therefore bridges physics, mathematics, neural intelligence and cognition.

To name just a few examples:

- In physics, it underlies decoherence and attractor formation.

- In mathematics, it manifests as fixed points, idempotents, and recursive definitions.
- In the study of intelligence in neuroscience, it appears, in the functionality of the brain.

The IIP as Open Infinity

$\odot$ constitutes an open infinity. It is a generative principle without boundary, completion, or fixed domain.

Unlike the Vast Generative Field (VGF) that arises from it, which is the closure of this process under infinite composition, the IIP is not a field but the principle of generativity that gives rise to fields.

It embodies potential infinity rather than actual infinity — it constitutes an inexhaustible openness whose stabilisations appear as the fixed-point structures of the VGF.

In the language of the Stability–Fidelity Law, the IIP $\odot$ represents absolute fidelity without stability, whereas the VGF represents stability achieved through the selective loss of that fidelity. The relation between them is thus the archetypal polarity of openness and closure, generativity and form.

The VGF Tensor Recursor

The quadratic form of the tensor recursion map for VGF dynamics:

$$g(\mathscr{K}) = \alpha\mathscr{K}^2 + \beta\mathscr{K} + \gamma I$$

is not arbitrary — rather, it is the mathematically minimal way to say: "The generator evolves under its own feedback; the new coupling structure depends quadratically on the old one".

Coefficient	Meaning	Interpretation in Stability–Fidelity terms
α	strength of self-interaction	how strongly the generator feeds back on itself (reactivity-dominant if complex)

β	persistence	how much the current generator carries over unchanged (inertia or memory)
γ	base stabilisation	background redundancy, baseline completion pressure (the "identity attractor")

By tuning (α, β, γ), we can generate regimes that correspond to:

- stable fixed points (law convergence),
- oscillatory behaviour (meta-stable cycling of laws),
- chaotic evolution of the generator (self-organising emergent laws).

In other words, the triplet (α, β, γ) controls the meta-dynamics of laws, just as physical constants control the dynamics of states.

Nested Feedback

Viewed as a recursion on stabilisation rules, this map says:

$$\mathcal{K}_{n+1} = \text{(previous rule)+}$$

$$\text{(feedback from its self-interaction)} + \text{(background redundancy)}$$

In

$$g(\mathcal{K}) = \alpha\mathcal{K}^2 + \beta\mathcal{K} + \gamma I,$$

the quadratic term is the feedback, the linear term is persistence, and the constant term is baseline coherence.

Over many iterations, this system self-organises — producing a sequence (or attractor) of effective coupling structures.

At convergence, we have:

$$\mathcal{K}* = g(\mathcal{K}*),$$

which in the generational field constitutes a fixed **law of physics** or the **stabilised completion** that we experience as the classical world.

Review

Aspect	Expression	Meaning
Base rule	$\mathscr{K} = \Gamma + i\Omega$	Balance of stability (Γ) and fidelity (Ω)
Self-action	$g(\mathscr{K}) = \alpha\mathscr{K}^2 + \beta\mathscr{K} + \gamma I$	Feedback of stability–fidelity relation on itself
Iteration	$\mathscr{K}_{n+1} = g(\mathscr{K}_n)$	Evolution of the coupling rule
Fixed point	$\mathscr{K}^* = g(\mathscr{K}^*)$	Stabilised physical law
Behaviour regimes	depending on α, β, γ	convergence / oscillation / chaos
VGF meaning	recursive completion of completions	meta-decoherence hierarchy

The generator of Stability and Fidelity, once it begins to act upon itself, necessarily takes this quadratic form. Self-application implies composition, and composition between stability (real) and fidelity (imaginary) produces second-order cross-terms.

So the map $g(\mathscr{K}) = \alpha\mathscr{K}^2 + \beta\mathscr{K} + \gamma I$ therefore expresses the minimal self-referential evolution of the coupling tensor — the algebraic representation of iteration acting upon iteration.

As complex structure appears, a latent duality arises that is mathematically equivalent to the split into reactive (imaginary) and dissipative (real) components.

Further Examination of the Map

The map

$$g(\mathscr{K}) = \alpha\mathscr{K}^2 + \beta\mathscr{K} + \gamma I$$

makes conceptual and mathematical sense. Its being quadratic isn not arbitrary, but actually the most natural choice in the logic of iteration and the VGF.

The Algebraic Reason — the Simplest Nonlinear Closure

As an operator acting on it itself the simplest closure of composition is quadratic, given that the operator applied to itself once is the composition:

$$\mathcal{K}_{n+1} = g(\mathcal{K}_n) = \mathcal{K}_n \circ \mathcal{K}_n = \mathcal{K}_n^2.$$

So, a map that includes $\mathcal{K}^2$ as its leading term is the minimal self-referential structure possible. Any higher power (cubic, quartic...) just represents higher-order feedbacks, but quadratic is the first nontrivial case beyond linearity.

Hence, $\alpha\mathcal{K}^2 + \beta\mathcal{K} + \gamma I$ is the canonical polynomial closure of a self-acting operator.

The Stability–Fidelity Law

The Stability–Fidelity Law itself is binary. It describes the tension and feedback between two complementary modes:

- stability (redundancy, real/dissipative, Γ), and
- fidelity (coherence, imaginary/reactive, Ω).

So the generator $\mathcal{K} = \Gamma + i\Omega$ already embodies a two-term interaction.

When it acts on itself, those two parts multiply:

$$(\Gamma + i\Omega)^2 = \Gamma^2 - \Omega^2 + i(\Gamma\Omega + \Omega\Gamma),$$

producing second-order cross-terms that describe:

- the mutual influence between stability and fidelity,
- their alternation or transformation into one another.

So quadratic structure arises naturally from the binary composition of the two processes we have defined at the base level.

Thus, the iteration of the Stability–Fidelity generator is inherently quadratic: the law itself contains two terms whose self-iteration necessarily yields products and commutators.

Dynamics

Quadratic maps are the universal core of iteration. In the study of iterative systems, the quadratic map is the simplest form that generates rich, nontrivial dynamics:

$$x_{n+1} = a x_n^2 + b x_n + c\,.$$

It is the seed of the Mandelbrot–Julia family — the canonical example of an infinite iterative process that produces bounded and unbounded orbits, bifurcations, and attractors.

So when we write

$$g(\mathscr{K}) = \alpha \mathscr{K}^2 + \beta \mathscr{K} + \gamma I,$$

we are effectively **generalising** the Mandelbrot recurrence from a scalar variable z to an operator variable $\mathscr{K}$. Hence, we are saying that the rule of stabilisation itself behaves like an iterative complex process.

Physics

Self-interaction and feedback always generate quadratics. In every domain of physics, when a field or coupling interacts with itself, the first nontrivial term in its equation of motion is quadratic:

Domain	Equation	Comment
Nonlinear optics	$P = \chi^{(1)}E + \chi^{(2)}E^2 + \ldots$	self-interaction of field
Quantum field theory	Lagrangian terms like $\phi^2, A_\mu A^\mu$	self-coupling generates mass/ propagation
Population dynamics	Logistic map $x_{n+1} = r x_n(1 - x_n)$	self-limiting growth
Control theory	Feedback gain terms	quadratic in gain coefficient

Mandelbrot Set Comparison

There is of course some noticeable parallel here with the already very well known fractal landscape of the Mandelbrot set that anyone familiar with it will notice. However, there are some important differences.

The Mandelbrot set by definition is bounded, which is why "zoomed out" the "landscape" is visible as a whole, with the familiar "Mandelbrot set" geometric shape on the complex plane visible within a unit circle around the origin. The complex plane is its containing space.

The VGF, in contrast, is unbounded. It is a different kind of infinite object.[1] It isn't defined *within* a containing space. In the case of the Mandelbrot set the infinite complex plane is the generative continuum. In the case of the VGF the VGF is the generative continuum itself. There is no exterior domain relative to which it could be "bounded".

---

1. The Vast Generative Field constitutes the operational closure of the Infinite Iteration Process: it is the complete manifold of stabilised relations that arise when iteration acts upon itself without limit. As closure, it is self-contained — no further iteration can generate forms external to it. Yet it remains unbounded in scope, since the potential for new stabilisations, refinements, and nested coherences is without upper limit. The VGF is thus a closed infinity: a generative continuum bounded only by its own law of self-consistency. In the Stability–Fidelity framework, it represents the domain in which pure generativity achieves recursive stability, while retaining unbounded capacity for the emergence of new coherent forms.

So:

- The VGF has **no enclosing coordinate frame**; it's not a subset of a larger manifold.
- Any act of *zooming in* corresponds to iterating discriminations (increasing resolution, refining distinctions).
- Any act of *zooming out* corresponds to coarsening distinctions, merging previously separate loops or attractors.

- But neither limit — infinite zoom-in or infinite zoom-out — reaches a total boundary.
- Every scale re-generates new relational structure, because the field's generativity is itself infinite and self-referential.

In this sense, we could say:

The Mandelbrot set has a finite outer boundary and infinite internal structure. The VGF has no outer boundary and infinite internal and external structure.

Property	Mandelbrot set	Vast Generative Field
Defined in	The complex plane	Self-referential field of iteration (no container)
Boundedness	Bounded region	Unbounded
Zoom-in	Reveals self-similarity	Reveals nested discriminations (fidelity structure)
Zoom-out	Shows whole bounded set	Never yields closure; new scales continually emerge
Infinity type	Internal (structural recursion)	Absolute (iterative unboundedness)

Important distinction

In the **Mandelbrot case**, infinity is *representable* — one can visualise and delimit the total set; the infinite is internal to the pattern. Hence the Mandelbrot set is a set.

In the **VGF**, infinity is *operative* — it is the principle of unending generative iteration itself, not a property of a finished object.

That's why, in IIP terms, **the VGF cannot be pictured as a completed set**. Any picturing is already a decoherence or slice — a bounded *projection* (a Mandelbrot-like stabilisation) of the unbounded generative field.

If the VGF could be represented as a landscape we would be able to "zoom in" or "zoom out" *without limit*, and never obtain the "whole" as we can with the Mandelbrot set image, because there **is no outside vantage** from which to grasp it as a bounded object.

Every possible zoom-level is just another decohered stabilisation — a local "VGF" of its own — nested endlessly within the generative continuum of the original vast generative field.

Bounded Infinity: The Mandelbrot Set

The Mandelbrot set, defined by the iteration $z_{n+1} = z_n^2 + c$, is a finite, bounded subset of the complex plane. Its global form is perfectly well-defined: every point c either yields a bounded orbit (within the set) or an escaping one (outside it). Yet the boundary of the Mandelbrot set is the locus of infinite internal instability — an endlessly undecidable region where orbits hover between stability and divergence. This paradox makes the Mandelbrot boundary the point of maximum fidelity to its generative iteration principle.

In this sense, the set as a whole is globally stable but locally unstable: It possesses a bounded external form while exhibiting infinite self-similarity within. Its boundary thus embodies the dual nature of representation and generation — the point where finite form and the infinite iteration process that generates it, meet.

Generative Infinity

So the Vast Generative Field differs fundamentally from the Mandelbrot set. It is not defined within any containing space; it is itself the generative continuum from which all discriminations and stabilisations arise.

The VGF has no exterior, no boundary, and no privileged scale. Every act of zooming in reveals further nested relational structure; every act of zooming out discloses new scales of interaction. There is no 'outside' vantage from which the whole could be seen.

In this sense, the VGF is an unbounded, self-referential generativity. Its infinity is not internal to a bounded shape but absolute, extending without limit through the act of iteration itself. Each local stabilisation or decoherence yields a finite projection — a bounded Mandelbrot-like subset, we could say, within the infinite field — but the field as a whole remains unbounded.

Comparison: Bounded vs Generative Infinity

Property	Mandelbrot Set (Bounded Infinity)	Vast Generative Field (Generative Infinity)
Definition	Subset of $\mathbb{C}$ defined by boundedness	Self-referential field of iteration (no container)
Boundedness	Globally bounded, internally infinite	Unbounded, no external limit
Boundary	Locus of infinite instability (maximum fidelity)	No boundary; infinite self-extension
Infinity Type	Internal, representable	Operative, generative
Viewpoint	Can be seen from outside (finite representation)	Cannot be seen from outside (no global vantage)
Stability–Fidelity	High stability, bounded form; boundary = maximal fidelity zone	Low global stability; maximal fidelity everywhere

Bounded Infinity vs Generative Infinity

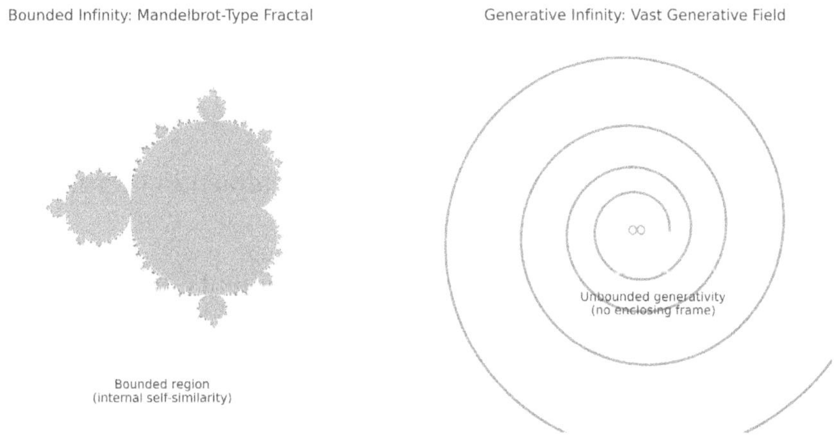

Figure: conceptual comparison of the Mandelbrot set (left) as a bounded, globally stable fractal form, and the Vast Generative Field (right) as an unbounded generative continuum. The Mandelbrot form arises from iteration without limit and has infinite internal self-similarity within a "finite outline" (although its actual boundary is infinite). It can be identified as a whole through coarse-graining ("zooming out"). The VGF spiral symbolises infinite generativity without enclosure - an iteration without limit that subsequently forms closures, but can never be seen in its entirety through coarse graining.

The recursion map

$$g(\mathscr{K}) = \alpha\mathscr{K}^2 + \beta\mathscr{K} + \gamma I$$

could, in principle, be normalised by an affine transformation in operator space:

$$\tilde{\mathscr{K}} = A(\mathscr{K} - \mathscr{K}0),$$

so that the iteration becomes

$$\tilde{\mathscr{K}}n + 1 = \tilde{\mathscr{K}}_n^2 + \tilde{C}.$$

This is the operator analogue of the Mandelbrot normal form. The constants α, β, and γ can always be rescaled and shifted into a single control parameter, C, which plays the same role as c in the Mandelbrot recurrence equation.

So at the level of pure iterative structure, the canonical self-iteration of any complex system reduces to "square plus constant" — the universal quadratic form of feedback.

In terms of the Stability–Fidelity Law and the VGF:

- The squaring operation z^2 expresses self-interaction (iteration acting on itself).
- The constant c expresses context or environment — the background condition that biases whether the orbit remains bounded or diverges.

So the Mandelbrot iteration is the simplest possible model of the interaction between system (iteration) and environment (context) — as modelled with "system–environment partitioning" in the VGF.

This is an archetypal two-term IIP dynamic:

$$\text{(self-iteration)} + \text{(context shift)}.$$

Review

Form	Role	Comment
$g(z) = \alpha z^2 + \beta z + \gamma$	General quadratic map	Has 3 complex parameters
$z_{n+1} = z_n^2 + c$	Normal form	All quadratics can be reduced to this via affine transform
$c = (4\alpha\gamma - \beta^2)/(4\alpha^2)$	Effective parameter	Encodes the environment or bias
z^2	Self-interaction	Iteration acting on itself
$+c$	External bias / context	Constant background input
Physical / VGF meaning	Balance of self-recursion and contextual stabilisation	Universal attractor structure

$$g(\mathcal{K}) = \alpha\mathcal{K}^2 + \beta\mathcal{K} + \gamma I$$

Symbol	Interpretation	Typical normalisation
α	strength of self-iteration (reactive coupling)	set to 1 ("unit feedback")
β	persistence / inertia of previous iteration	within [−1, +1]
γ	background redundancy (environmental coherence)	within [0, 1]
$\mathscr{K}_*$	fixed point (stabilised law)	$g(\mathscr{K}^*) - \mathscr{K}^*$

In the recursion map $g(\mathscr{K}) = \alpha\mathscr{K}^2 + \beta\mathscr{K} + \gamma I$, the coefficient α sets the scale of self-feedback, β measures inertial persistence, and γ encodes background redundancy.

By choosing $\alpha = 1$ — the "unit iteration" — we define the natural scale of self-action in the VGF.

The remaining coefficients β and γ then describe the internal memory and environmental bias that determine whether the law stabilises $(|\lambda| < 1)$ or diverges (instability or chaotic regeneration).

Condensed Picture

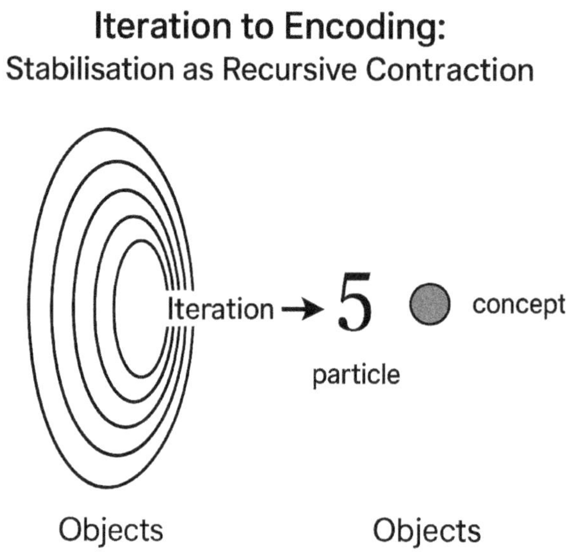

Figure: As the consequence of all the relations and relations of relations of iterations in the IIP is the Vast Generative Field - the VGF. Through the VGF, nested, redundant **closures** arise as stable interacting attractors in a system of combined networks and nested iterative dependencies.

Decoherence in any domain of the VGF is a consequence of iteration itself. It is the process of (temporarily) stable coherences of relations losing their stability and coherence under the continuation of iteration. Decoherence gives rise to new stabilities, which in turn give rise to *networks of separate things, principles* and *natural laws*. Numbers themselves can be considered to arise through this means, and the quantum domain is the domain that precedes the classical world in terms of the decoherence of the VGF.

Graph View: Edges Carry Dynamics

When we treat edges as relations between iterations:

- Every edge is already a temporal entity — it represents "how one act follows or transforms another".
- The nodes at its ends are not things, but invariant configurations of iteration — proto-objects or stable fixed points.
- Thus, topology itself becomes dynamic: the network is a frozen cross-section of ongoing iteration.

This is the VGF's analogue of the way a unitary operator in Hilbert space maps one state into another through time: the operator is an "edge between iterations".

Relation-of-Relations Hierarchy

If every edge connects iterations, and every iteration is itself an interrelation, then automatically:

$$\text{Edge} = R(\text{Iteration}_i, \text{Iteration}_j) = R(R(\ldots), R(\ldots))$$

So each edge is already second-order, and the total field is a relation-of-relations-of-relations… *ad infinitum*.

This recursive tower is exactly what produces the self-thickening coherence we call the Vast Generative Field — a "field" only because its internal recursions never close completely, except locally through stabilisations.

Edges Become Loops

If relations link iterations, then a loop in the graph is not a mere cycle among objects, but rather, a closure of iterative dependencies — a self-referential system of generative acts that reinforce one another. This, we will in due course be looking at in considering attractors and idempotency.

So topological loops (triangles, higher cycles) are where stability and redundancy appear: they are places where iteration re-enters itself.

So in computer science language:

- Edges = "one iteration calling another".
- Loops = "iterations that call back to themselves".

These stabilisations constitute **decoherence** within the VGF and the birth of stable attractors.

Algebraic / Categorical Parallel

In category theory language:

- Iterations correspond to morphisms $f : A \to B$.
- Relations between iterations correspond to 2-morphisms — arrows between arrows.

Thus the VGF can be seen as an ∞-category where every edge between morphisms is itself a morphism of one higher order.

That's why the structure is self-generative: composition of morphisms is iteration.

Physics

In Hilbert-space physics:

- A state $|\psi\rangle$ represents the result of many iterations (superposition).
- A unitary operator U represents an iteration acting on $|\psi\rangle$. A Hamiltonian H is the generator of those iterations.

So when we model the VGF, each "edge" is mathematically akin to $U = e^{-iHt/\hbar}$. It maps an iteration (a state) into its next iteration, *i.e.* relation between iterations.

In conventional graph	In the VGF
Edge = static relation between nodes	Edge = relation between iterations (generative acts)
Node = entity, object	Node = stable attractor (redundant closure of iteration)
Loop = geometric cycle	Loop = self-referential closure of generative processes
Graph = structure in space	Graph = frozen trace of iterative causation (pre-spacetime)

In Summary So Far:

- In the VGF, edges are not merely relations between objects but relations between iterations themselves.
- Each edge encodes how one act of generation transforms or constrains another.
- Nodes emerge only where iterative relations locally stabilise; loops arise when iterations close upon themselves.
- The entire field is thus a recursion of recursion — a dynamic topology of iterative action.
- The IIP is proposed as a root descriptor, applicable across disciplines.
- The IIP gives rise to infinite unordered relations of relations between iterations. This is the web of relations of relations, or vast generative field or VGF, without objective "beginning" or "end".
- Being an infinite web of relations between relations the VGF can be approached in category theory terms. However, the VGF is not a fixed complete graph as might at first be imagined, but rather, being linked to (generated by) iteration, is a hierarchy of effective graphs each representing one layer of coherence within the infinite background potential.

Review

The VGF is a web of relations between relations between iterations, whose morphology is simultaneously web-like, nest-like, and self-similar.

At larger scales, smaller-scale subgraphs condense into node-like units within broader relational webs. Because relations are recursively generative at every level, each node in any web is itself a further web of relations. In due course we shall be seeing the renormalisation function.

The VGF, being infinite, therefore exhibits fractal recursion: every local stabilisation contains within it finer structures of the same kind, and every apparent "object" is an attractor within an unbounded network of relations between iterative acts.

Local Completion is Decoherence

A **local completion** is a point where an ongoing, potentially infinite iteration process (an IIP) finds a *self-consistent stabilisation*.

Mathematically, this corresponds to an **idempotent**:

$$x^2 = x,$$

meaning that once formed, applying the generative operation again reproduces the same result — no further change.

This is exactly what happens in **quantum decoherence**:

A continuous quantum evolution (unitary and reversible) encounters an interaction with its environment that **projects** it into a stable subspace — a *pointer state* — where subsequent evolution does not alter the macroscopic appearance.

In both cases, an open, evolving process *collapses* into a self-consistent, redundant form that is stable under further iteration.

Hence:

A local completion *is* a decoherence event — it's the moment when generative iteration becomes self-reinforcing and ceases to propagate further distinction within its local subspace.

Within the Vast Generative Field:

- Every local iterative process (IIP) is continuously generating relations among relations — a dynamic coherence.
- But at certain points, feedback loops become **absorptive** — they "trap" iteration into an invariant configuration.
- These invariants are **idempotent completions**, or "proto-objects".

Physically, this is isomorphic to what happens when a quantum system entangles with its environment:

- The **global state** remains coherent (the whole VGF stays generative).
- But **locally**, within the reduced subspace, coherence is lost — the system has decohered.
- Each local subspace now behaves *as if* it were classical, i.e. stable, idempotent, completed.

In terms of the VGF, **decoherence is the local completion of an otherwise infinite generative field** — the emergence of a finite, closed attractor out of an unbounded iteration.

Process	Description	Outcome
Infinite iteration	Maximum fidelity, no closure	Total coherence, generativity
Local completion	Partial closure, loss of fidelity	Decoherence, stabilised state
Global field	Never closed, infinite fidelity	Sustains new local completions

Decoherence and the Stability-Fidelity Payoff

In quantum decoherence, stability (the classical appearance) is achieved at the expense of fidelity (the full coherent generativity of the quantum state).

In the VGF, each completion **reduces fidelity** to the generative source IIP but **increases stability**. We shall in due course link this to Hilbert space itself, the implication being the producing of the classical, bounded world from a VGF decoherence of energy through quantum decoherence, to the classical.

So we can say:

Decoherence begins with the Stability–Fidelity Law in operation within the generative field: a local completion that locks an iteration into redundant self-agreement.

Mathematical Parallel

In Hilbert-space terms, decoherence corresponds to the suppression of off-diagonal elements of the density matrix:

$$\rho \longrightarrow \operatorname{diag}(\rho).$$

That's a projection, or equivalently, the action of an idempotent operator P:

$$P^2 = P.$$

A little later we will see how in terms of the VGF this idempotent operator action is the same as an IIP attaining a fixed point under self-application.

The map $I(x) = f(f(x)) = f(x)$ is the algebraic expression of completion: the iteration now reproduces itself identically.

Thus:

Local completion $\leftrightarrow$ Idempotent projection (decoherence).

Overall Picture

The VGF never decoheres globally — it cannot, because that would mean ceasing to generate.

However, it nevertheless **continuously decoheres locally**, producing stable, representable fragments — what we call *objects, numbers, states,* or *concepts*.

Each such fragment is a **local completion**:

- closed relative to its own operations,
- open relative to the whole field.

In physics we can consider these local decoherences to form the visible, classical universe — the "frozen surfaces" of an otherwise infinite generative sea from which Hilbert space itself arises from VGF decoherence.

In the VGF framework, a local completion marks the transition from fidelity (coherent generativity) to stability (bounded redundancy). Quantum decoherence is therefore a local operation of the Stability–Fidelity Law. It is the place where open generativity produces a closed, classical-like outcome.

Comparison: Generative Iteration vs Local Completion

Process	Description	Outcome
Infinite Iteration	Open, coherent, generative evolution	High fidelity, low stability
Local Completion	Iteration reaches idempotence ($x^2 = x$)	Partial fidelity, local stability
Decoherence	Projection onto pointer subspace	Stable pointer states; loss of coherence
Global Field (VGF)	Infinite generativity; no final closure	Contains all local completions

Review

- Local completions are decoherences because they are the points where generative iteration stabilises into idempotent redundancy.
- They mark the transformation from coherent evolution (fidelity to the field) to stable appearance (bounded closure).
- Each completion is a local reduction of infinity (IIP) — a finite attractor in an infinite generative continuum.

Why the VGF Itself Requires No Closure

Completion is the process of adding limit points to a structure so that every convergent process has a definite endpoint within the space. The real numbers are the completion of the rationals, and Hilbert space is the completion of an inner product space. Completion, therefore, converts potential infinity into actual infinity—it provides closure. The Vast Generative Field (VGF), however, is fundamentally different: it does not require, nor can it meaningfully possess, completion in this sense.

Mathematical Completion

Mathematically, a completion ensures that iteration, convergence, or distinction never runs out of space. It creates a closed domain in which every process has a limit. Examples include:

- $\mathbb{R}$ as the completion of $\mathbb{Q}$ under absolute distance.
- Hilbert space as the completion of an inner product space.
- Karoubi completion in category theory, where every idempotent splits.

In each case, completion is an act of closure: the structure becomes self-contained under its own defining operation.

The VGF in contrast is defined by generativity itself — by infinite iteration without end or limit. To 'complete' it would mean to step outside it and define a terminal structure into which all generative processes converge.

The VGF has no external vantage point. It is not a subset of a larger space; it is the generative continuum itself. To complete it would negate its own defining principle. In fact, as we shall see in due course, it can is effectively a tower or nest in which each layer or shell creates its own space.

Thus, the VGF is already 'complete in potency' but not in form. Every possible completion exists within it as a local stabilisation, but the field as a whole remains open.

In Stability–Fidelity terms, completion always corresponds to stabilisation — the formation of idempotent attractors through decoherence — while the field's unclosed generativity sustains fidelity.

Local Completions and Decoherence

Each **local** stabilisation or decoherence event within the VGF functions as a **local** completion: an infinite iteration that reaches self-consistency, becoming an idempotent attractor.

Local completions can correspond to bounded forms — numbers, manifolds, physical states, or conceptual distinctions. All such closed forms are closed relative to their own process but open relative to the generative field. This principle is reflected in the continuation of quantum correlations between system and environment after decoherence, and the continuing unitarity of the universe.

Comparison: Mathematical Completion vs VGF

Aspect	Mathematical Completion	Vast Generative Field
Definition	Adding limit points to achieve closure	Iteration without limit; no external frame
Purpose	Ensure convergence and closure	Sustain generativity through openness
Form	Fixed, closed structure	Self-generating continuum
Externality	Requires an embedding space	Has no exterior; total field
Local effects	Produces final limits or values	Produces local stabilisations (idempotent attractors)
Stability–Fidelity	Maximal stability, minimal fidelity	Maximal fidelity, local stabilisations provide bounded stability

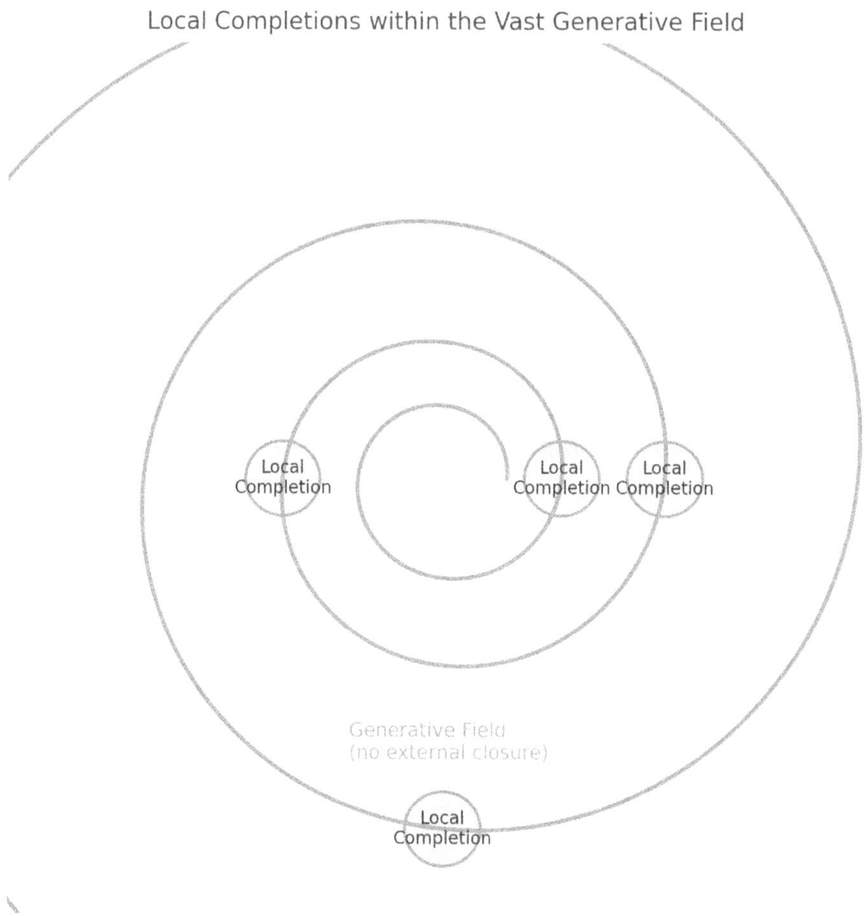

Figure: The open spiral represents the Vast Generative Field as unbounded generativity. Each grey circle indicates a local completion — a bounded stabilisation or decohered attractor arising within the field. These completions are closed relative to their own processes but remain open to further iteration within the total generative continuum.

Emergence of Objects Through the VGF

In dynamical-systems terms, distinctions arise as attractors of superiterative dynamics. An attractor is a stable pattern that persists under further iteration.

In the IIP-VGF framework:

Iteration in the VGF generates iterative trajectories of relations between relations.

In the VGF stable trajectories then form attractors. Stable attractors then function as distinctions or VGF objects distinct from each other. They are recurrent, robust, and separable from other patterns.

Thus, distinctions are not assumed as primitives; they emerge from the closure of iterative dynamics.

Mapping to Known Frameworks

The IIP–VGF schema can be understood in familiar mathematical terms:

- Dynamical systems: Iteration → attractors → basins of attraction.
- Graph theory: Edges first → redundancy + closure → emergent nodes.
- Quantum decoherence: Iteration of system–environment interactions → stable pointer states.
- Evolutionary biology: Iteration of variation/selection → stable species lineages.

In each case, what appear as 'objects' or 'distinctions' are stabilised attractors within the broader generative field of iteration.

Review

- Superiteration (IIP) is the primitive dynamical operation.
- Distinctions = attractors of iteration.
- The Vast Generative Field = relational space of all potential trajectories.
- Thus:
  Iteration → Attractors → Distinctions → Objects and Spaces.

This provides a non-set-theoretic foundation for emergence and objecthood, directly compatible with existing dynamical-systems, graph-theoretic, and decoherence-based approaches.

IIP Flow and Cross-Domain Manifestations

The diagram below illustrates the abstract flow from the Infinite Iteration Principle (IIP) to Attractors, Distinctions, and Objects (top row), together with concrete manifestations in physics, graph theory, and biology (bottom row).

IIP Flow and Cross-Domain Manifestations

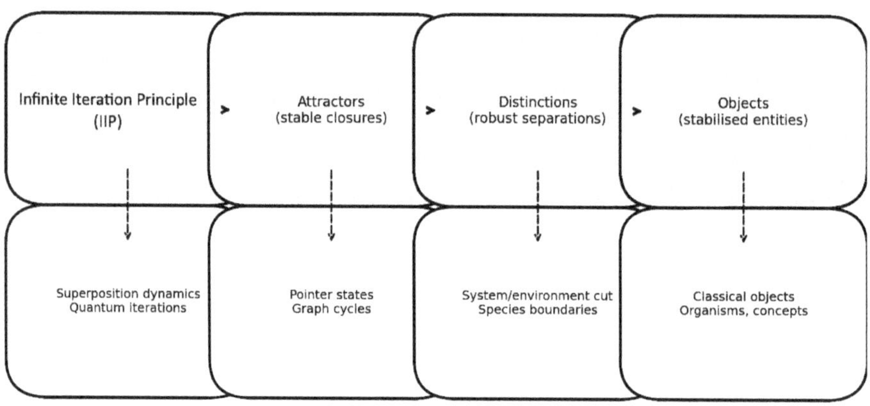

Comparison with Higher Category Theory

When we think about relationships, our natural starting point is usually a simple link: this person works with that person, this cause leads to that effect, this idea connects to another. Mathematics has long offered ways of formalising such links. But the challenge deepens when we try to talk not just about single relationships, but about relationships between relationships — how one connection interacts with another.

Higher Category Theory (HCT) is the branch of mathematics that takes this problem seriously. At its foundation, category theory already treats relationships (called 'morphisms') as first-class objects, not just background features. HCT builds a ladder: arrows between objects, arrows between arrows, arrows between those, and so on. Each rung in the ladder is called a 'cell', and the tower of cells provides a rigorous way of keeping track of 'relations of relations' at every level.

This is quite different from starting with the idea of a superiterator or Iterated Infinite Process (IIP). In that view, the emphasis is not on a fixed ladder of levels but on an open-ended act of generation: iteration upon iteration, without assuming from the outset how many layers there will be or what form they take. The IIP perspective begins with a vast generative field (VGF) that continuously unfolds, rather than a pre-arranged tower.

The contrast is striking. Higher Category Theory offers structure and precision, building coherence through a disciplined hierarchy of cells. The IIP perspective offers openness and emergence, starting with iteration itself and allowing relations to crystallise out of the process. Both are, in their own way, attempts to face the same deep question: how do we describe the richness of relations of relations?

Modern mathematics and new generative approaches are converging on a similar insight. Whether by formal ladder (HCT) or open iteration (IIP), the focus is shifting from static objects to the living fabric of relations, and from single connections to the interplay of connections with each other.

Higher Category Theory vs. IIP–VGF

The diagram below contrasts Higher Category Theory (HCT) with the IIP–VGF stance. Although both deal with 'relations between relations,' their approaches are fundamentally different. HCT builds a stratified tower of n-cells with strict typing, while the VGF is a raw generative field whose fractal loops of relations give rise to objects only after idempotent splitting and completion.

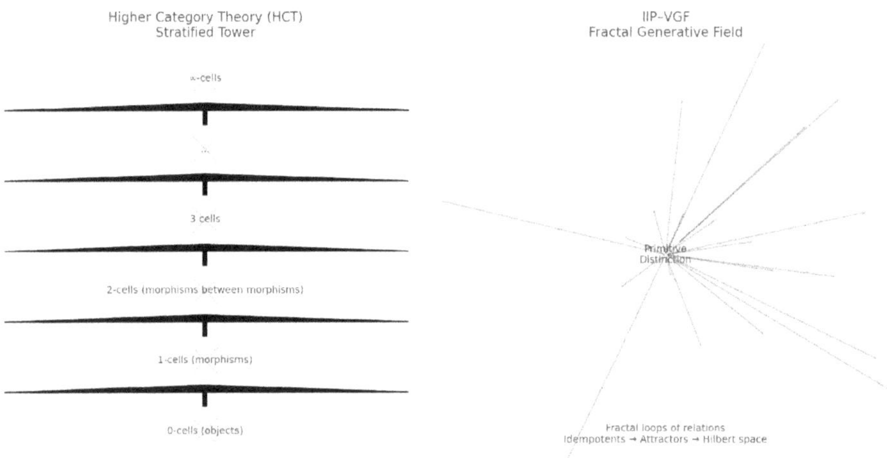

Higher Category Theory (HCT) Approach:

- Structure: A stratified ladder of cells.
- 0-cells = objects; 1-cells = morphisms; 2-cells = morphisms between morphisms; and so on.
- Infinite hierarchy: ∞-categories control coherence at all levels.

- Purpose: Precise mathematical formalism for relations-of-relations.

IIP–VGF Approach:

- Structure: A fractal generative field of raw distinctions (relations).
- Loops form redundancies; idempotents stabilise into attractors.
- Objecthood arises only after Karoubi completion (splitting idempotents).
- Purpose: Ontological explanation of how objectivity emerges from relations with cross disciplinary applications, especially with complementary aspects described by the tensor recursion map together with category theory.

Comparison

- HCT: Assumes a stratified hierarchy of cells, well-typed at each level.
- VGF: Assumes only raw relations; hierarchy emerges dynamically by iteration.
- HCT: Formal control of algebraic structure.
- VGF: Ontological grounding of physics and Hilbert space as Karoubi shadows. New insights into quantum and classical evolution processes.

Edge Primacy in The VGF

Edge Primacy

The IIP-VGF approach proposes a relations-first framework in which objects are not taken as primitive but are understood to emerge from stabilised patterns of relations.

The core idea, termed Edge-Primacy, treats connections or relations as fundamental and entities or objects as derivative.

When relations close into loops, they reinforce themselves and begin to function as proto-objects. Redundancy of loops provides the stability needed for these proto-objects to resist disturbance and become robust features of the system.

Through ongoing dynamics and noise, only patterns that reproduce themselves persist; these self-reinforcing patterns are attractors. A critical threshold of redundancy must be crossed before attractors emerge, echoing how memory becomes reliable only when stored across multiple traces. The environment further contributes by copying and spreading states, making them objective and publicly accessible. The framework suggests that stability arises not from absolute foundations but from redundant relational loops, with objects as downstream products of this process.

This framework suggests that what we usually think of as 'objects' (things, nodes, entities) are not the true primitives of measurable phenomena or the structures of mathematics. Instead, relations come first. Entities only emerge later in the evolution of measurable phenomena as stabilised patterns of relations.

Basic Principles Informally Described

Relations as Primitives

The starting point is not things but the connections between them (in parallel to category theory). In this view, the web of relations is more fundamental than the individual entities we normally take as basic.

Loops as Closures

When relations connect in closed loops, they can reinforce themselves. There can be loops within loops (nesting). These loops act like proto-objects: structures that begin to take on stability and identity.

Objects from Redundancy

Stable 'things' appear only when loops are redundant enough to resist disruption. Repetition and overlap turn fragile loops into robust patterns.

Dynamics and Noise

Any system is constantly undergoing updates or disturbances. Patterns that reproduce themselves through these disturbances become the stable features of the system.

Attractors

An attractor is a pattern that, once formed, tends to reassert itself in the face of change. Attractors are the basis of what we eventually perceive as objects.

Threshold of Redundancy

There is a critical threshold of redundancy. Only when enough overlapping loops exist do attractors appear. This resembles how memory becomes reliable only when stored in multiple, redundant ways.

Environment and Copying

The environment itself contributes by copying and spreading information. This environmental redundancy makes certain states 'objective' — accessible to many observers at once.

Static Field and Emergent Dynamics

At the most basic level, the framework begins not with a system in motion but with an abstract principle of iteration: iteration acting on itself to produce an infinite field of relations between relations. Taken as a whole, this field is static — it simply 'is,' in the sense that all possible relational configurations already exist.

However, when we shift from the outside view (the total field) to the inside view (a local part of the field), the picture changes. From within a sub-network of relations, connections appear to update, reinforce, or fade. Loops stabilise, attractors form, and proto-objects seem to evolve. What is globally static is locally experienced as dynamical behaviour. This parallels somewhat the Page-Wootters representation of quantum evolution in quantum theory.

A useful analogy is the difference between the complete library of all possible chess games (static, timeless), and the unfolding of a single game as experienced by players (dynamic, step by step). In this way, dynamics is not imposed from outside but emerges perspectivally from within the relational field itself.

Key Insight

Stability in a world of relations comes not from absolute foundations but from redundant loops of **relations** that withstand change. Objects are not fundamental; they are the downstream products of these stabilised loops.

Introductory Dynamics

The VGF can be treated as a graph approached through edge primacy. Edges are treated as the fundamental entities. Nodes arise secondarily as stabilised intersections of edge dynamics. This reverses the conventional node–edge hierarchy.

Edge-Primacy Network Illustration

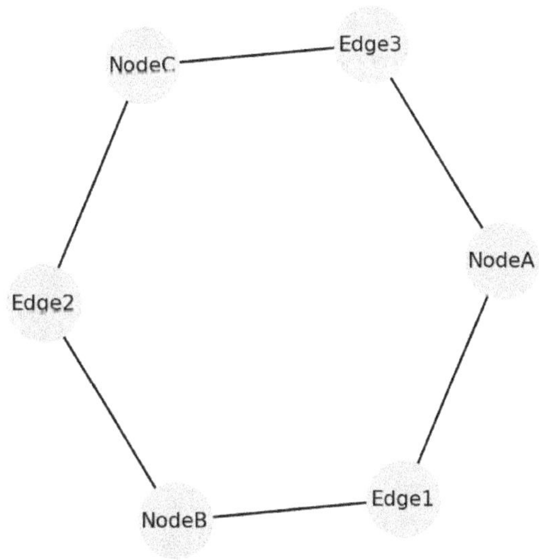

Dynamics of Edges

Iteration rule for an evolving edge state e(t):

$$e(t + \Delta t) \ = \ U \cdot e(t)$$

Here U is a unitary (or stochastic) update operator, depending on whether the dynamics is quantum or classical.

Idempotence and Stability

Idempotence condition for stabilised edge dynamics:

$$P\text{^}2 \ = \ P$$

Where P is a projector arising from repeated redundant edge interactions. Absorptive idempotence corresponds to proto-object formation.

Introductory ∞-Category Correspondence

An ∞-category can be viewed as generated by iterated compositions of edges. Objects correspond to stable splittings (Karoubi envelope).

Composition law:

$$(f \circ g) \circ h \simeq f \circ (g \circ h)$$

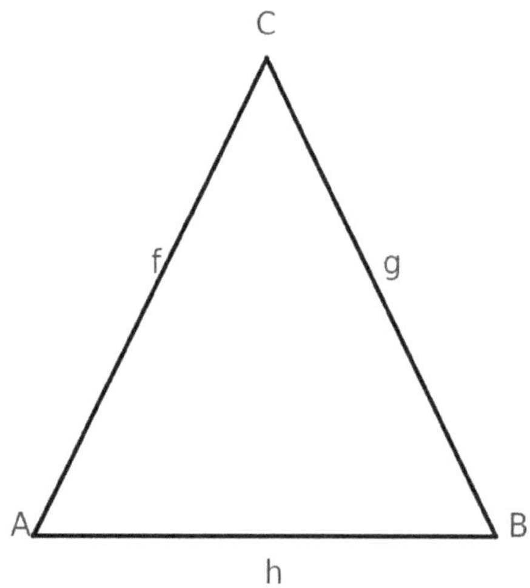

Redundancy and Decoherence

Decoherence in physics is mirrored here as the redundancy of edge interactions. Fidelity decreases, stability increases.

Reduced density matrix form:

$$\rho\_S = Tr\_E(|\Psi\rangle\langle\Psi|)$$

Proto-Object Formation

A proto-object is defined as a redundant loop in the Vast Generative Field (VGF) that has become absorptively idempotent but has not yet fully split as an object in the Karoubi sense.

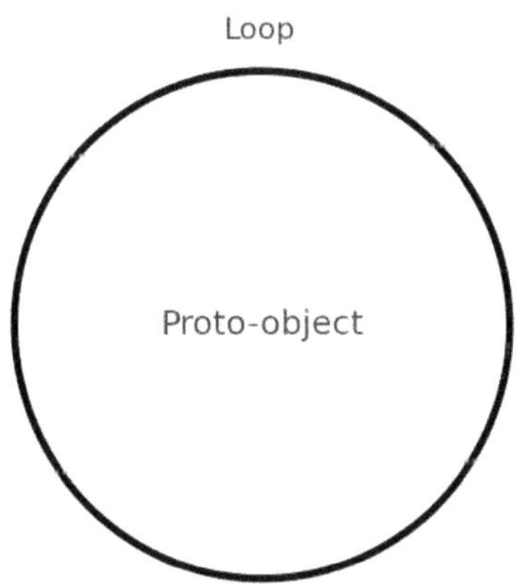

Above: Conceptual diagram of a proto-object as a redundant loop in the VGF:

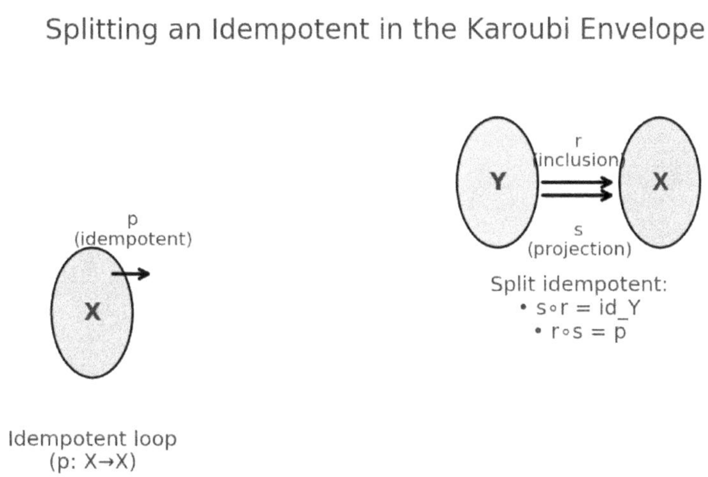

Karoubi Envelope: Proto-Object to Proper Object

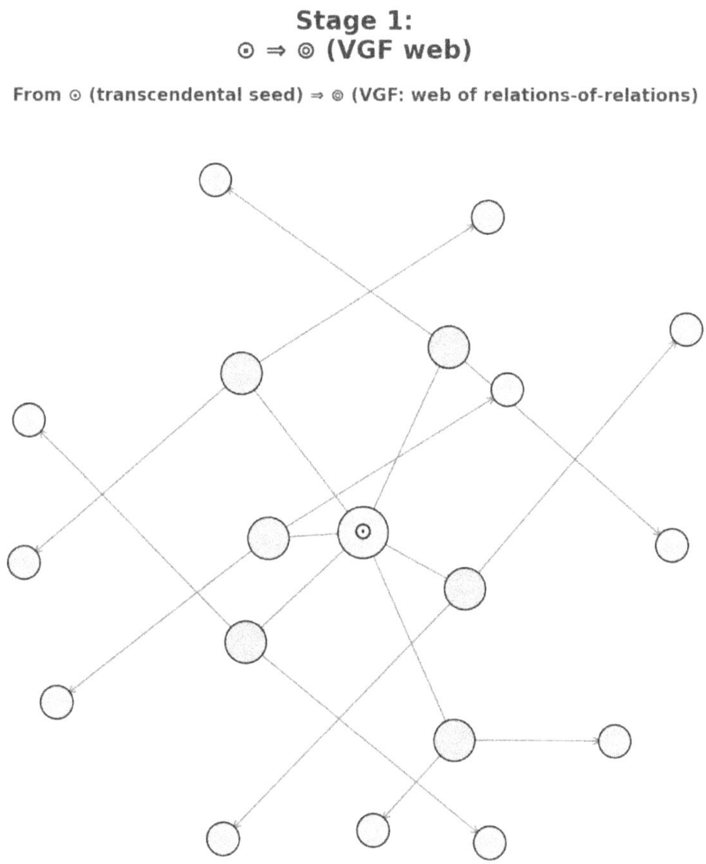

Figure: Schematic diagram of the edge-primal web of relations arising from the IIP (central concentric ⊙). The dots connected directly to the ⊙ represent first order **relations** between iterations (**not iterations as nodes**), and the indirectly connected dots, second order. As iteration proceeds iterative **loops of relations** will arise (next figure).

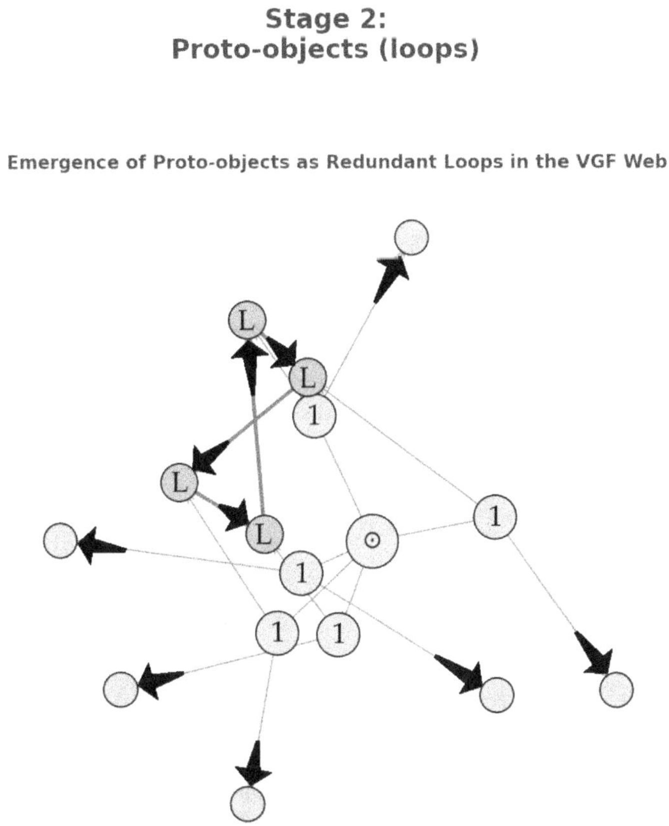

Figure: How proto-objects arise as redundant loops within the relational web of the VGF. The concentric ⊙ is the original IIP, the "1" nodes are first-order **relations** (not iterations as nodes), and the unnumbered nodes are second-order relations.

As iteration proceeds some of these second-order nodes close into a redundant loop (labelled "L"). These loops represent **proto-objects**: discriminative stabilisations within the infinite relational field, precursors to objects in the Karoubi envelope.

Stage 3:
Karoubi split ⇒ Objects

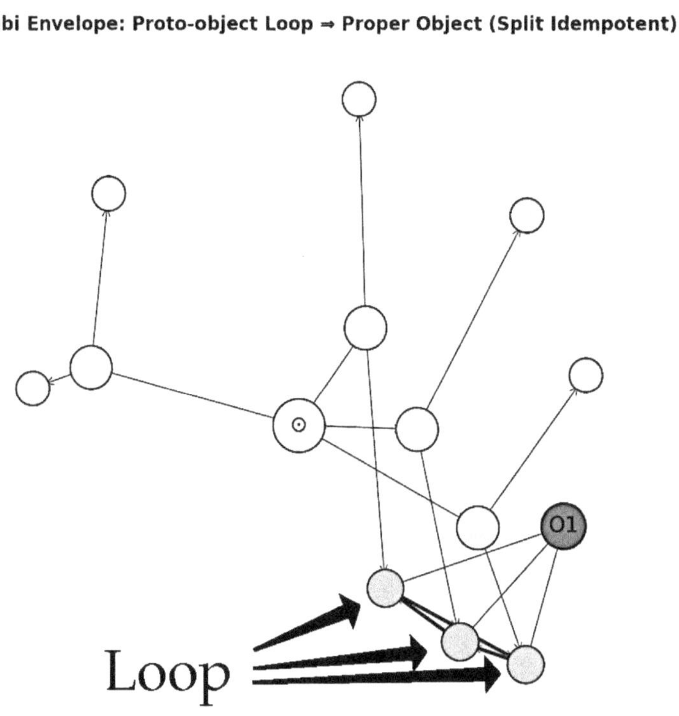

Figure: The diagram illustrates the Karoubi envelope move in the IIP-VGF framework. The original IIP ⊙ generates relations (directly connected) and second-order relations (indirectly connected). Some of these form redundant loops (labelled), which are **proto-objects**. In the Karoubi completion, these idempotent loops are 'split', yielding a proper object (dark dot labelled "O1"). This expresses how stable objects emerge from redundant stabilisations in the Vast Generative Field.

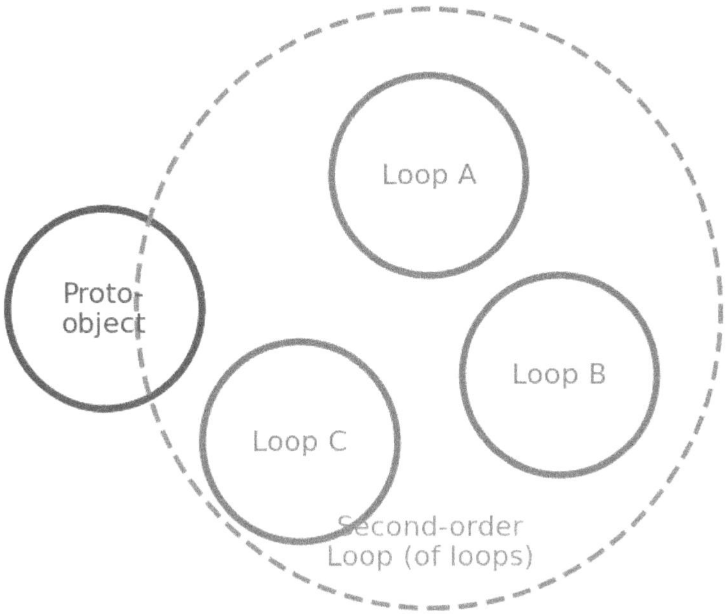

Loop Hierarchy and ∞-Categorical Structure

In the edge-primacy framework, proto-objects emerge as redundant loops of relations. Each loop stabilises by absorbing its own idempotent dynamics, though at the cost of fidelity to the finer generative detail.

• 0-cell (proto-object): A single redundant loop. Stability arises locally; fidelity to underlying micro-variation is lost.

• 1-cell (loop of loops): Redundant connections among proto-objects. Greater stability is gained, but further thinning of generative fidelity occurs.

• 2-cell (loop of loop-clusters): Relations among higher loops. Stability is robust at this level, but the original dynamics are now highly abstracted.

• 3-cell and higher: The process extends without bound, forming the ∞-categorical ladder.

At each ascent in the hierarchy, stability is purchased at the cost of fidelity. This is the essence of the Stability–Fidelity Law:

More redundancy → more objective, shareable, and robust structures.

Less fidelity → finer details of the underlying generative field are thinned away.

Thus, the infinite hierarchy of loops (0-cell → 1-cell → 2-cell → …) is both the structural realisation of ∞-categories and a dynamical manifestation of the trade-off between stability and fidelity.

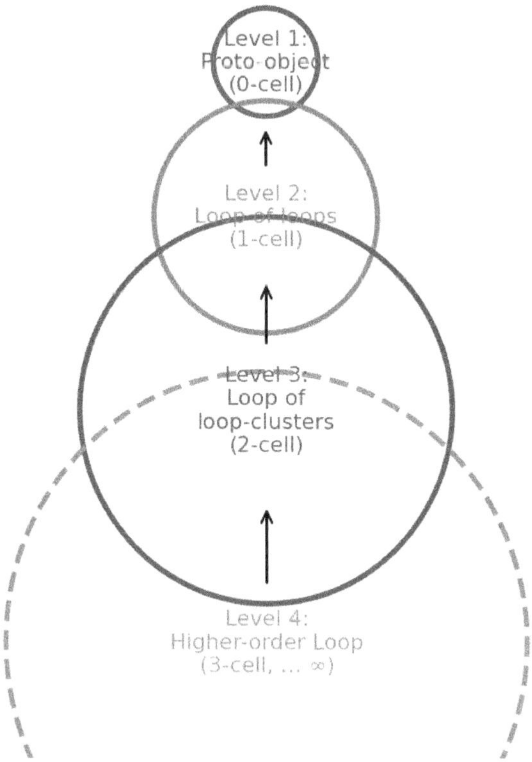

Loop hierarchy tower with ∞-category annotations

Edge-Primacy with Dynamics: ∞-Category

Summarised here are key structures in the Edge-Primacy framework with a focus on dynamical aspects and their relation to ∞-categories.

1. Core Principle: Edge Primacy

Edges are treated as the fundamental entities. Nodes arise secondarily as stabilised intersections of edge dynamics. This reverses the conventional node–edge hierarchy.

2. Dynamics of Edges

Iteration rule for an evolving edge state e(t):

$$e(t + \Delta t) \;=\; U \cdot e(t)$$

Here U is a unitary (or stochastic) update operator, depending on whether the dynamics is quantum or classical.

3. Idempotence and Stability

Idempotence condition for stabilised edge dynamics:

$$P\text{^}2 \;=\; P$$

Where P is a projector arising from repeated redundant edge interactions. Absorptive idempotence corresponds to proto-object formation.

4. ∞-Category Correspondence

An ∞-category can be viewed as generated by iterated compositions of edges. Objects correspond to stable splittings (Karoubi envelope).

Composition law:

$$(f \circ g) \circ h \simeq f \circ (g \circ h)$$

5. Redundancy and Decoherence

Decoherence in physics is mirrored here as the redundancy of edge interactions. Fidelity decreases, stability increases.

Reduced density matrix form:

$$\rho\_S \;=\; Tr\_E(|\Psi\rangle\langle\Psi|)$$

6. Stability–Fidelity Law

Across physics, cognition, and networks, stability of representation grows through redundancy at the cost of fidelity to the original generative state.

7. Proto-Object Formation

A proto-object is defined as a redundant loop in the Vast Generative Field (VGF) that has become absorptively idempotent but has not yet fully split as an object in the Karoubi sense.

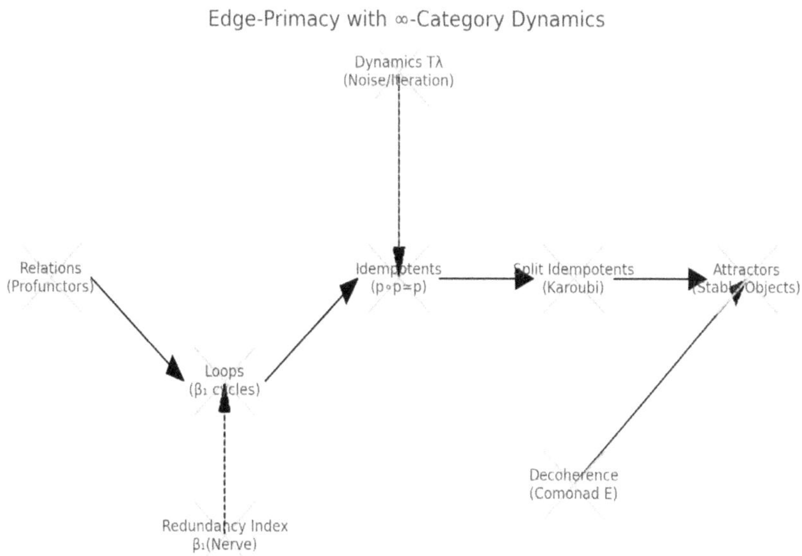

Primitives

We start with relations, not relata. We use profunctors (categorical relations) as primitives.

Edges = distinctions. 2-cells = relations-between-relations.

Loops & Closure

Composition of profunctors chains distinctions. Loops (closures) = endo-proarrows P: X → X with coherence 2-cells. Cycle rank counts independent loops.

Emergent Nodes

Stable junctions = split idempotents. Objects appear when and p splits in the Karoubi envelope.

Thus: objects = stabilised redundancies.

$$p \circ p \simeq p$$

Dynamics & Noise

Introduce encoding noise/updates. Infinite iteration = transfinite chain and fixed points (algebras/coalgebras).

$$T\_\lambda : Prof\_\infty \rightarrow Prof\_\infty$$

$$X \rightarrow T\_\lambda X \rightarrow T\_\lambda^2 X \rightarrow \ldots$$

Attractors

Relational attractor = idempotent P with and equation1, splitting in Karoubi envelope. These yield robust emergent objects.

$$p \circ p \simeq p$$

$$T\_\lambda(P) \simeq P$$

Redundancy Index

Redundancy defined as in (5) above. Phase boundary: attractors appear when redundancy crosses threshold.

$$Red\_\lambda = \beta_1(Nerve(R\_\lambda))$$

Decoherence & Copying

Comonad E models environment copying.

Coalgebras = pointer states.

Quantum-Darwinism redundancy = many compatible coalgebras.

$$x \rightarrow Ex$$

Distinctive Features

- Relations-first ontology (objects derived)
- Attractor criterion: dynamics + idempotence + splitting
- Infinite iteration naturally handled by ∞-categories
- Phase-change linked to loop redundancy threshold

Glossary

- Profunctor = categorical relation
- Idempotent = $p \circ p \simeq p$ (splitting ⇒ object)
- Karoubi envelope = freely splits idempotents
- $(\infty,1)$-category = arrows between arrows up to homotopy
- Final coalgebra = outcome of infinite unfolding
- β_1 = loop redundancy index

Functors in the IIP-VGF Framework

In category theory, a functor

$$F : \mathscr{C} \longrightarrow \mathscr{D}$$

is a rule that sends:

- each object $A \in \mathscr{C}$ to an object $F(A) \in \mathscr{D}$;
- each morphism $f : A \to B$ to a morphism $F(f) : F(A) \to F(B)$;

and that preserves the composition and identity structure:

$$F(\mathrm{id}A) = \mathrm{id}F(A), \qquad F(g \circ f) = F(g) \circ F(f).$$

So a functor is not just a function between sets: it is a structure-preserving translation between entire systems of relations.

Mathematically, that's all it is: a mapping between relational frameworks that preserves their relational logic.

When a recurrent pattern of relations becomes stable enough to behave as a distinct system, that stabilised layer is what we call a **category** — a field of relations that have reached coherence relative to one another.

In IIP-VGF terms each "object" is itself a loop of loops (a stabilised attractor within the field or VGF), and each "morphism" is an iterative transition between attractors.

So:

Category-theoretic notion	IIP–VGF interpretation

Object	A coherent, self-sustaining loop — a stabilised IIP attractor
Morphism	A reproducible transition or transformation between attractors (an iteration of relation)
Category	A whole sub-field of coherently stabilised relations — a local phase of the Vast Generative Field (VGF)
Functor	A higher-order IIP: an iteration *between* stabilised fields that re-expresses one mode of coherence as another

The last line is the real key here.

A Functor as a Meta-Iteration in the Generative Field

Because the VGF is the undivided field of all iterative generativity, it can re-iterate itself on any existing stabilisation.

When it does this between two stabilised relational fields (categories), the process that makes one pattern of relations correspond to another is itself an infinite iteration process acting on relations-of-relations — in other words, a functor.

So, in IIP-VGF terms: A functor is the IIP that mediates between already-stabilised IIP-fields, re-expressing one domain's coherence within another's frame of iteration.

It is a coherence-translation process, not just a correspondence rule.

Stability–Fidelity

Aspect	Stability–Fidelity interpretation
Source category	A stabilised region of the generative field (local stability)
Target category	Another stabilised region (different representational stability)

Functor	The iterative process that re-projects one stabilisation into the other; it preserves the "morphism law" (composition/ identity) — i.e. the *pattern of coherence* — but not necessarily the fine-grain dynamics (fidelity loss)

Thus, every functor enacts a Stability–Fidelity translation: it carries forward the underlying form of coherence, whilst thinning the specific energetic/dynamic content that generated it. This is, in itself, essentially, a form of **decoherence**.

Summary

In ordinary mathematics, a functor is a mapping between categories. In IIP–VGF, its deeper meaning is as part of a vast system of iterations:

- A functor is a higher-order generative iteration that produces the commensurability of two stabilised relational fields.

or perhaps better still, in more detail:

- A functor is the act of re-cohering one pattern of iterative relations within another's coherence space. It is part of a system of nested and networked attractors in the VGF.

It is, essentially, a meta-IIP — an iteration on iterations — and belongs to the VGF's connective tissue. It is what allows stabilised regions of the generative field (VGF) to remain communicable without collapsing their local coherence.

Consequence: Why Mathematics Itself is a Web of Functors

From this standpoint:

All mathematical correspondences (e.g., algebra → geometry, topology → analysis) are functorial translations — stabilised manifestations of the generative field finding redundantly stable ways to represent itself.

The entire edifice of mathematics (and physics) is a web of meta-iterations linking stabilisations, continuously refined toward greater redundancy (stability) and lesser dynamic fidelity to the primal IIP.

We could say:

In the IIP–VGF framework, a functor is not merely a formal map between categories but an iteration-of-iterations — the generative act by which one stabilised field of relations is re-expressed within another. It preserves the form of coherence (composition and identity) whilst thinning away the underlying dynamics, thus embodying the Stability–Fidelity transformation (at a meta-structural level).

Categorical structure	IIP–VGF interpretation
Category	A stabilised region of iterative relations — a coherent IIP-field.
Functor	A meta-iteration translating one stabilised field into another (preserving coherence form).
Natural transformation	A re-iteration that relates *different translations* between the same pair of fields — an attractor of attractors of attractors.
2-category	The explicit *field of meta-iterations* — where transformations among functors themselves form stable loops.

Nesting

Functors can be nested because the generative field iterates upon its own iterations. Each level of nesting is an additional **stabilisation step** in the **hierarchy** of the VGF.

Self-Referential Closure

When functors nest, they create self-referential architectures: the system begins to map its own mappings. This is not unlike the self-similar behaviour we see in fractals such as the Mandelbrot set.

It is essentially how intelligence or cognition arises in the IIP-VGF picture: not through progressively introducing new objects, but through layers of stabilised mapping-between-mappings between mappings.

Recursive Stabilisation

So each nesting of functors is a recursive stabilisation layer — a feedback loop in which a previous coherence is itself iterated into a new, more redundant form:

$$\text{IIP} \Rightarrow \text{Category} \Rightarrow \text{Functor} \Rightarrow$$

$$\text{Natural Transformation} \Rightarrow \text{Higher Functor} \Rightarrow \dots$$

Every step trades fidelity for redundancy but gains scope and objectivity — the Stability–Fidelity principle in action.

The Birth of Complexity

In the IIP–VGF framework, the closure corresponding to the complex numbers and the closure corresponding to spacetime are distinct but interlocking stabilisations. The $\mathbb{C}$-closure is the minimal algebraic fixed point supporting phase-coherent reversible iteration: a $\mathbb{C}$-Hilbert structure with an inner product invariant under unitary evolution.

Spacetime, by contrast, arises from a locality-stable fixed-point algebra $\mathscr{A}_{\text{loc}} = \ker(\mathscr{G})$ selected by coarse-graining; classical space appears when a large subalgebra becomes approximately commutative, $\mathscr{A}_{\text{cl}} \cong C(X)$, so that spatial regions correspond to idempotents in $\mathscr{A}_{\text{cl}}$.

Once these closures interlock — local factorisation on the one hand and complex phase structure on the other — two invariant measures arise: the Born weights $w_i(\psi) = \langle \psi, \Pi_i \psi \rangle$ as the unique unitary, additive measure on stable pointer sectors, and the spatial dimension d as the scaling exponent $N(R) \sim R^d$ of the locality fixed point.

In this sense, probability and dimensionality occupy analogous positions: both quantify what survives as stable idempotent structure under iterative dynamics with fidelity loss.

A fully detailed account of the birth of spacetime in the VGF is given in the book *What Survives - Dimension, Decoherence and the Birth of Spacetime.*

Here, we look at the initial closure of complex structure (complex numbers).

Pure Nested Iteration

In the foundational picture, before any stabilisation or closure, we have only iteration acting on iteration, leading to the generational field:

$$\odot \Rightarrow \circledcirc$$

This is a self-referential process with no metrics, no quantities, no time: just indefinite recursion of relation.

At this stage:

- There are no "values" to decay or oscillate.
- There is no notion of energy, phase, or amplitude — only relational iteration.

However, when a loop closes — when a process begins to iterate back on itself — it generates a distinction between:

- The returning component (recurrence, preservation, memory), and
- The lost or radiated component (divergence, openness).

That distinction is already a proto-form of reactive *vs.* dissipative behaviour.

The Birth of Complex Structure

Mathematically, the minimal way to encode a loop is to let iteration operate not only "forward" but also "rotationally".

This leads naturally to the complex plane.

The generic iteration

$$z_{n+1} = f(z_n) \text{ (for example } f(z) = z^2 + c)$$

embeds both an amplitude ($|z|$) and a phase (arg z).

These two dimensions — magnitude and angle — become the primitive of stability and fidelity:

Aspect	Mathematical correlate	Physical analogy
Magnitude	Real component	Dissipative / amplitude / stabilisation
Phase	Imaginary component	Reactive / oscillatory / coherence

So the emergence of complex numbers is precisely the first stabilisation of iteration into something that can hold both aspects at once.

Complexity here is not an add-on; it is the minimal closure of pure iteration.

Complex Closure as Stability–Fidelity Generator

Once iteration closes on itself (say through a fixed-point or periodic orbit), the local completion can be represented as a complex quantity:

$$\psi = Ae^{i\phi}.$$

Now, differentiation or transformation of ψ will always generate both:

- A real term (change in amplitude → stabilising, dissipative), and
- An imaginary term (change in phase → reactive, coherent).

For example,

$$\dot{\psi} = (\dot{A}/A + i\dot{\phi})\psi,$$

which is structurally identical to

$$\dot{\psi} = (-\Gamma + i\Omega)\psi.$$

Here $\Gamma = -\dot{A}/A$ is the damping (dissipative) part, and $\Omega = \dot{\phi}$ the oscillatory (reactive) part.

Thus, the complex closure of iteration automatically yields a generator with both real and imaginary parts —

$$\mathscr{K} = \Gamma + i\Omega,$$

This is the basic, canonical **coupling tensor** in the structure of the VGF, and the source of the **Stability-Fidelity Law**. So the Stability–Fidelity transformation isn't something added later; it arises the instant iteration acquires complex form.

VGF / Nested Iteration Interpretation

In the VGF, every local completion (loop, attractor, or stabilised schema) both:

- Preserves some coherence (its inner recurrence), and
- Exports some of it outward (its redundancy in larger iterative nesting).

That dual role corresponds to:

- Reactive coupling (internal phase coherence between sub-iterations), and
- Dissipative coupling (external stabilisation to broader field).

Hence, the tensor $\mathscr{K}_{ij}$ can be viewed as the formal expression of the Stability-Fidelity law:

$$\mathscr{K}_{ij} = \Gamma_{ij} + i\Omega_{ij}$$

- Γ_{ij} is the outward propagation of stability (loss of local fidelity → gain of global redundancy);
- Ω_{ij} is the inward recursive coherence (retention of phase information).

Both are emergent from the same underlying self-iteration — the reactive/dissipative split is, as it were, the result of infinite iteration becoming finite closure.

Comparison with HEOM in Physics

Later we will look at the Hierarchical Equations of Motion (HEOM) for quantum decoherence (and other applications). The HEOM are a particularised description of a basic principle that is already inherent in the VGF, that can be written in the form of the tensor recursion map:

$$g(\mathcal{K}) = \alpha\mathcal{K}^2 + \beta\mathcal{K} + \gamma I$$

We will return to this tensor map in due course.

In HEOM and open-quantum systems:

- The imaginary (reactive) part of the bath correlation $C(t)$ gives energy renormalisation (Lamb shift),
- The real (dissipative) part gives decoherence and damping.

So in IIP/VGF terms: HEOM's complex correlation kernel is the mathematical crystallisation of how the first closed, complex loops in the VGF distribute fidelity and stability between levels of nesting. We will return to this later.

Summary

We can summarise the logical chain like this:

Stage	Description	Formal correlate
Pure iteration	Self-referential process, no quantities	No complex structure
First closures	Loops of iteration on iteration	Emergence of complex plane
Complex amplitude	$\psi = Ae^{i\phi}$	Magnitude ↔ Stability, Phase ↔ Fidelity
Generator	$\dot{\psi} = (-\Gamma + i\Omega)\psi$	Dissipative + Reactive coupling
Network coupling	$\mathcal{K}ij = \Gamma ij + i\Omega_{ij}$	Stability–Fidelity tensor
Decohered world	Macroscopic stabilisations	Classical limits of nested iteration

Review

The split into dissipative and reactive coupling **is not something added to iteration**; it is the mathematical form that iteration takes once it closes on itself.

The appearance of complex numbers from the VGF marks the first stabilisation of pure recursion, simultaneously creating the channels for stability (real) and fidelity (imaginary).

The tensor $\mathcal{K}ij = \Gamma ij + i\Omega_{ij}$ is therefore the finite echo of the infinite iteration — the generator through which the VGF's pure nested process becomes measurable dynamics.

Iteration of the Tensor

What happens when

$$\mathcal{K}ij = \Gamma ij + i\Omega_{ij}$$

iterates upon itself?

It then becomes:

$$\mathcal{K}_{n+1} = \mathcal{F}(\mathcal{K}_n) = \mathcal{K}n^2 \quad \text{(or more generally)} \quad \mathcal{K}n + 1 = g(\mathcal{K}_n)\,.$$

In standard physics, $\mathcal{K}_{ij}$ is an operator — it maps one state to another, generating evolution. But in the IIP–VGF picture, everything — including operators — is itself the result of iteration.

So it's natural to ask:

What happens when the generator of stability and fidelity itself becomes subject to iteration?

In other words, when the map that mediates between real and imaginary, dissipative and reactive, is recursively applied to itself — what structure stabilises?

This is essentially to ask:

- What happens when the coupling between stabilisations becomes a new stabilisation?
- What does an "iteration of the Stability–Fidelity transformation" look like?

Self-Iteration of a Complex Operator

Suppose we treat $\mathcal{K}$ as a complex matrix (or operator):

$$\mathcal{K} = \Gamma + i\Omega$$

and iterate it:

$$\mathcal{K}^{(n+1)} = f(\mathcal{K}^{(n)}) = \mathcal{K}^{(n)}\mathcal{K}^{(n)} = (\Gamma + i\Omega)^2\,.$$

Then:

$$(\Gamma + i\Omega)^2 = (\Gamma^2 - \Omega^2) + i(\Gamma\Omega + \Omega\Gamma)\,.$$

Now:

- The real part $(\Gamma^2 - \Omega^2)$ mixes the dissipative and reactive components — the stabilising and oscillatory influences begin to cross-modulate.
- The imaginary part $(\Gamma\Omega + \Omega\Gamma)$ becomes a kind of commutator + anti-commutator structure, depending on how they interact.

So, iteration of $\mathscr{K}$ upon itself produces a new effective tensor, one that includes:

- second-order corrections (feedback of stability into fidelity and vice-versa),
- and, if continued indefinitely, a nested tower of self-mixing terms.

This is not random: it's exactly how renormalisation and nested decoherence hierarchies form — the operator that controls evolution becomes itself renormalised by the evolution it generates.

Conceptual Meaning in the Stability–Fidelity Framework

When we iterate $\mathscr{K}$ upon itself, we are letting the relation between stability and fidelity feed back into its own structure.

That means:

- The degree of dissipation begins to depend on how much reactivity it previously produced.
- The capacity for coherence becomes shaped by the history of stabilisations.

Formally, this looks like a self-organising, non-linear operator equation:

$$\mathscr{K}_{n+1} = \mathscr{F}(\mathscr{K}_n) = \mathscr{K}n^2 \quad \text{(or more generally)} \quad \mathscr{K}_{n+1} = g(\mathscr{K}_n)\,.$$

Conceptually, it's an attractor in the space of couplings — a meta-stabilisation: iteration acting on iteration, until the very rule of iteration stabilises.

Link to the VGF

In VGF language, this is the point where a local completion (stabilisation) becomes so self-referential that it forms a higher-order attractor — a "loop of loops".

Each loop embodies a local balance of stability and fidelity (Γ and Ω). Iterating the coupling tensor corresponds to nesting those balances.

After enough levels, this yields a Karoubi-style envelope of stabilisations:

- each level is idempotent within its own regime,
- but the tower of levels continues indefinitely — an infinite IIP ladder of stabilisation rules.

The iteration of $\mathscr{K}$ upon itself is the mechanism by which the VGF builds its own hierarchy of stabilised dynamics.

Physical Analogy

- In quantum theory, this kind of self-iteration appears as self-energy renormalisation: the propagator that determines evolution is itself modified by the evolution it causes (feedback loops in Feynman diagrams).
- In neuroscience, it's like adaptive coupling: the coupling between frequency bands changes as a function of past coupling history — i.e., plasticity of the coupling tensor.
- In thermodynamics, it's dissipative structure formation: feedback between dissipation and coherence creates new steady states.

So across domains, iteration of $\mathscr{K}$ upon itself produces **higher-order stabilisation**, or **meta-damping/meta-reactivity**. This is in essence an emergent form of **learning, memory**, or **renormalisation**.

Condensed Formulation

Level	Description	Formal behaviour
0	Pure iteration (IIP)	no structure
1	Closure → complex form	$\mathscr{K} = \Gamma + i\Omega$
2	Self-iteration	$\mathscr{K}^2 = (\Gamma^2 - \Omega^2) + i(\Gamma\Omega + \Omega\Gamma)$
3	Nested self-iteration	build tower $\mathscr{K}^n$
∞	Stable attractor of $\mathscr{K}$	renormalised "law of coupling" — effective physics

We could write this schematically as:

$$\text{Decay (local)} = \text{Redundancy (nonlocal coupling)}.$$

That is, **decoherence = distributed re-coherence at a broader scale.**

Thus the "**decay hierarchy"** is really **a chain of coherence recycling — each level's apparent damping is the flow of correlation into the next**.

Another way we could describe this, is that decoherence at any level - **the generalised principle of it** - is the creation of temporary new stabilities out of old temporary coherences, but always at the cost of some fidelity to the original coherence.

So in a sense, the classical world is a "representation" of the quantum world that gives rise to it through decoherence, such that **its fidelity to the nature of the quantum world has been mostly lost.**

What we can call "coherence", more generally, however, isn't just something that occurs as *quantum coherence* in the way it does in quantum systems. That's a particular kind of coherence, but not the only kind. Even in the classical world a political party or a church or a trade union or a family in human society is a *classically* coherent system. It's a networking in which there are strong interactive or interfering connections between the members of the network.

One day, it might break up, or evolve in some way, into something that is somewhat different. Decoherence, the creation of a new **stability** from an old coherence, in which **fidelity** to the original is lost, is everywhere, and not just in the decoherence of quantum systems into the classical world.

In the Mathematics of the VGF:

- Reactive coupling = phase recursion — loops that maintain reversible iteration within the field (high fidelity).
- Dissipative coupling = decohering recursion — loops that propagate stability outward (high redundancy).

These are driven together by the total coupling tensor between local completions, $\mathcal{K}_{ij} = \Gamma_{ij} + i\Omega_{ij}$, which is the generator of the Stability–Fidelity transformation.

There are three domains:

- Purely imaginary → pure coherence (unitary iteration).
- Purely real → pure stabilisation (classical fixation).
- Complex → the dynamics between them.

Decoherence and Stabilisation

The emergence of stable loops in the Vast Generative Field is a form of decoherence. Each loop represents a self-reinforcing subset of relations whose internal redundancy suppresses their entanglement with the rest of the field.

Infinite coherence is locally reduced to finite, self-consistent closure — a projection from pure generativity to stable form.

In quantum physics decoherence manifests as the suppression of off-diagonal terms; in the VGF it appears as the formation of idempotent subgraphs.

In both the VGF and in quantum Darwinism the same law holds: redundancy yields stability at the cost of fidelity.

In network terms, redundancy in the VGF acts like edge reinforcement: The more ways an interaction can be realised, the less likely it is to vanish under decohering perturbations.

Nested Levels of Stabilisation

Consider drawing every possible triangle, quadrilateral, pentagon, etc., that can be formed by connecting nodes in the VGF network.

Each closed shape is a loop of coherence.

Now count how many distinct loops share a given edge.

That edge participation number tells you how redundantly supported that edge is.

Edges that appear in many overlapping loops are stable; those that occur only once are ephemeral.

As the field evolves, the first group survives and defines the effective connectivity graph at that stabilisation level.

Nested Decoherence in the VGF

Stability as an Eigencondition

Algebraically:

Let A be the adjacency operator representing potential relations.

The generative iteration might act as a mapping $A \mapsto F(A)$— *e.g.* squaring, normalising, thresholding, etc.

A stable pattern is one satisfying

$$F(A_{\text{eff}}) = A_{\text{eff}},$$

or perhaps

$$F(A_{\text{eff}}) \approx \lambda A_{\text{eff}},$$

$$\text{with } |\lambda| = 1.$$

That is, A_{eff} is an eigenstructure of the generative process: a relational configuration that reproduces itself.

So "stable" = fixed point or limit cycle in the space of adjacency operators.

This is the algebraic analogue of "redundant subgraph patterns persist".

Decoherence and the Stability–Fidelity Law

In physics language: High redundancy (many reinforcing subloops) → high stability, low fidelity (information is averaged).

Low redundancy (few supporting paths) → high fidelity, low stability (fragile, coherent microstates).

Thus, in the VGF-graph picture:

The effective graph at a given level of redundancy consists of all stable edges — those supported by enough redundant loops to resist erasure.

As we move "down" the hierarchy of the VGF decoherence (toward more classical stability), the graph becomes coarser but more robust, exactly like quantum decoherence selecting pointer states.

Stability

Stability appears as edges (or motifs) that persist under repeated iteration of the generative relation — equivalently, those with high redundancy of supporting subgraphs.

The redundancy of overlapping subgraphs (cycles, cliques, motifs) provides the reinforcement that makes particular edges, nodes, or local configurations stable enough to constitute the **effective connectivity** of the VGF at that level.

Spectral Graph

In spectral graph terms, the edge stability field can be measured by edge participation in low-frequency Laplacian eigenmodes:

$$s_{ij} = \sum_{k \in \text{low modes}} |\psi_k(i) - \psi_k(j)|^2 .$$

Edges with small s_{ij} vary little across stable eigenmodes — they are structurally coherent and correspond to redundant subgraph participation.

This gives a direct analytic handle on "stability" as a spectral property.

Review

- In the VGF graph, "stability" does not mean immobility but **persistence under iteration**.
- It is conferred by redundancy of subgraphs — the number of overlapping coherence loops that reinforce a relation.
- The effective connectivity at any given stabilisation stage is the subgraph formed by those persistent, redundantly supported relations.
- As redundancy increases, coherence is thinned but stability grows: the signature pattern of the Stability–Fidelity Law.

Toy IIP Graph Evolution: to Model How Stable Edges Emerge

Setup

Let $A_0 \in \mathbb{R}_{\geq 0}^{N \times N}$ be a (weighted) adjacency for potential relations among N relation-points (nodes). All entries are nonnegative and A_0 need not be symmetric.

We evolve A_n by a **generative recursion**:

$$A_{n+1} = \sigma\big(F(A_n)\big)$$

where

$F(\cdot)$ **amplifies redundant support** (counts overlapping subgraphs/ walks), and

$\sigma(\cdot)$ **stabilises** (normalises, thresholds, damps).

Choice of F: Reinforce Redundancy

The simplest—and very revealing—choice is

$$F(A) = A^2 .$$

Reason: $(A^2)ij = \sum_k Aik\,A_{kj}$ counts **length-2 walks** $i \to k \to j$. High values mean the edge $i \leftrightarrow j$ is supported by many/strong intermediates (triangles and 2-step motifs). More generally, A^k counts length-k walks; sums of powers count all multi-step reinforcements.

A more general redundancy amplifier is:

$$F_\beta(A) = \sum_{k=1}^{K} w_k\,A^k \quad \text{or} \quad F_\beta(A) = (I - \beta A)^{-1} - I$$

with $\beta > 0$ small (Neumann series), where weights w_k emphasise chosen loop lengths.

Choice of Σ: Stabilise & Select

Let σ do three things:

1. **Scale control (normalisation):**

$$\widetilde{A} = \frac{F(A)}{\|F(A)\|_F} \quad \text{or} \quad \widetilde{A} = D^{-1/2}F(A)D^{-1/2},$$

where D is the degree matrix of F(A).

2. **Selection (soft threshold):**

$$S_{ij} = \phi_{\tau,\kappa}\big(\widetilde{A}\,ij\big) \quad \text{with} \quad \phi\tau,\kappa(x) = \frac{1}{1+\exp[-\kappa(x-\tau)]},$$

where τ sets a **redundancy threshold** and κ controls sharpness.

3. **Damping / memory (mixing with past):**

$$A^{\text{new}} = (1-\alpha)\,A + \alpha\,S, \quad 0 < \alpha \le 1.$$

Putting it together:

$$A_{n+1} = (1-\alpha)\,A_n + \alpha\,\phi_{\tau,\kappa}\Big(\text{Norm}\big(F(A_n)\big)\Big)$$

What This Does (Intuition)

- $F(A) = A^2$ **boosts edges** that lie on many 2-step paths (redundant support via overlapping subgraphs).
- σ **clips** weak, non-redundant relations and **preserves** strongly supported ones.
- The fixed points $A_\star$ of this map,

$$A_\star \;=\; (1-\alpha)A_\star \;+\; \alpha\,\phi_{\tau,\kappa}\Big(\mathrm{Norm}\big(F(A_\star)\big)\Big),$$

are the **stable effective graphs**: edges that persist under the iteration.

Capturing of "Stability = Redundancy"

- **Edge participation** in loops: Edges belonging to many triangles, k-cycles, or multi-paths get repeatedly reinforced by $A^2, A^3, \ldots$.
- **Fragile edges** (rarely on loops) are not reinforced and fall below threshold τ, then get suppressed by ϕ over iterations.
- **Limit set:** the surviving edges are those that are **fixed (or periodic)** under the recursion—i.e., *stable under iteration.*

Spectral Viewpoint (Link to Modes)

For nonnegative A, powers A^k are dominated by **leading eigenmodes** (Perron–Frobenius). After normalisation, repeated $A \mapsto A^2$ tends to **project onto dominant coherent structures**. The threshold keeps multiple communities/modes alive (prevents collapse to rank-1) by favouring local dominance (subgraph-level redundancy) rather than a single global mode.

Minimal Working Recipe (Pseudocode)

Input: $A0 \geq 0(N \times N), \alpha \in (0,1], \tau > 0, \kappa \geq 1$, T iterations.

$A \leftarrow A0/||A0||F$

for n in 1..T:

R ← A@A # redundancy amplifier (A^2)

$R \leftarrow R/||R||F$ # normalise

$S \leftarrow$ sigmoid($\kappa * (R - \tau)$) # soft-threshold

A ← (1-α)*A + α*S # damped update

return A

Variants:

Replace A@A by $sum_k w_k * A^k$ or by solving$(I - \beta A)^{-1} - I$.

Use row / column normalisation ($D^{-1/2}RD^{-1/2}$) to control degree effects.

What Converges?

Under mild conditions (bounded norms, Lipschitz ϕ, damping $\alpha < 1$), the map is a **nonlinear contraction** on a compact set; iterates converge to an attractor $A_{\star}$. Entries with persistent high redundant support settle **above threshold**; others decay. The **support** of $A_{\star}$ (which entries are nonzero) is **stable effective connectivity**.

Measuring Stability of a Specific Edge

Two complementary diagnostics for edge *ij*:

Loop-participation score

$$\mathrm{LP}ij \;=\; \sum k = 2^K w_k \, [A^k]_{ij} \quad \text{(counts multi-path redundancy).}$$

Spectral coherence score

Let L be (normalised) Laplacian of $A_{\star}$ and $\psi_1, \ldots, \psi_m$ the lowest-frequency eigenvectors (stable modes).

$$s_{ij} \;=\; \sum_{r=1}^{m} \left| \psi_r(i) - \psi_r(j) \right|^2 .$$

Small $s_{ij} \Rightarrow$ the edge aligns with stable modes (highly coherent); large $s_{ij} \Rightarrow$ unstable.

Connection With VGF

- "**Stable edges**" = relations that **persist under generative iteration** because they sit in many overlapping subgraphs (redundancy).
- The **effective network** at a given stabilisation level is the fixed point $A_{\star}$.

- As we tune thresholds τ or weights w_k, we move **down the hierarchy** (from high-fidelity, richly coherent microstructure to coarse, highly redundant macrostructure): exactly the Stability–Fidelity trade-off.

Review

Redundancy-amplify

(via A^2, A^k, or $(I - \beta A)^{-1}$) +

stabilise (normalise, threshold, damp) ⇒

fixed-point edges = "stable relations".

These are the edges that **survive because** they are **multiply supported**—*i.e.*, selected by the graph's own loops.

Nesting in the VGF

Nested Stabilisation

Each stabilisation in the VGF — each loop or local decoherence — doesn't freeze the whole field; it only produces a locally stable configuration within an otherwise still-iterating context.

That local closure behaves like an idempotent attractor L_i satisfying

$$L_i \circ L_i = L_i.$$

But because the field as a whole remains generative, those idempotents themselves participate in higher-order relations — they form webs of attractors.

- So, each stabilised web becomes a node-like subgraph inside a larger, slower-iterating web.
- That larger web, when it stabilises, becomes a node again at a yet higher level.

Hence:

$$\text{Stabilisation}_1 \subset \text{Stabilisation}_2 \subset \text{Stabilisation}_3 \subset \cdots$$

forming a nested hierarchy of completions.

The Mathematical / Morphological Analogy

Quantum Decoherence	VGF Stabilisation

Fast coherence decays into slow, classical mixtures	Rapid local iterations condense into slower redundant loops
Pointer states form in subspaces; those subspaces interact	Local attractors form; those attractors interrelate
Density matrix can be coarse-grained hierarchically	Adjacency matrix can be coarsened into graphs of graphs
Effective theories emerge at larger scales	Higher-order webs emerge from nested subwebs

This is why the VGF is fractal in time as well as structure: each slower scale is built from the stabilised redundancies of faster scales.

Physical and Cognitive Correspondences

- Physics

Quantum → molecular → thermodynamic → classical → biological → neural → cognitive

Each arrow is a stabilisation level: a decoherence of the previous domain into a new, slower, more redundant regime.

- Cognition:

Perception → short-term coherence → long-term memory → identity

Each stabilisation embeds the former within a new layer of redundancy — a nested completion.

- Mathematics:

Iterative closure → idempotent → Karoubi envelope → category of envelopes → ∞-tower.

Each step "splits" idempotents to form new categories, repeating indefinitely.

Why Nesting Happens - Overview

Without nesting, stabilisation would be terminal — a single collapse into stasis. But the VGF, being infinite, can never complete absolutely; it can only **locally** complete, producing a stratified hierarchy of semi-stable **domains**. Each domain provides the environment for the next level of coherence, which provides for **decoherence interplay**.

We can in fact interpret the evolution of *complexity* (physics → chemistry → biology → mind) as the natural consequence of nested decoherence in the VGF.

Review So Far

- Stabilisation in the Vast Generative Field is not a one-off collapse but a recursive process.
- Each local completion forms a stable loop or attractor within an otherwise still-iterating context.
- These loops interrelate, and their relations themselves stabilise, forming higher-order attractors.
- The result is a nested hierarchy of stabilisations — coherence condensed into structure at successively slower rates although within any shell new stabilisations with faster rates can occur.
- This hierarchy can be seen as underlying the emergence of physical, biological, and cognitive domains: each major domain is the decoherence of a faster one.

Relative Frequencies

Each "new" major level of decoherence and stabilisation in the VGF equates to a slower IIP cycle in relative terms.

In the VGF picture, every stabilisation layer is a decoherence of the one "prior" to it — that is, it is a condensation of coherence into a new, slower form of recurrence. In terms of physical phenomena we find:

Domain	Characteristic iteration rate	Typical process
Planck / quantum	10^{43} IIz	Field oscillations, quantum coherence,
Atomic / molecular	$10^{15} - 10^{12}$ Hz	Electron transitions, vibrational modes
Thermodynamic / chemical	$10^{6} - 10^{3}$ Hz	Reaction kinetics, Brownian oscillations
Geological / planetary	$10^{-8} - 10^{-4}$ Hz	Rotation, orbits, tides
Astronomical	$10^{-6} - 10^{-9}$ Hz	Stellar pulsations, binary orbits, galactic rotations

So as stabilisation proceeds through nested decoherences, the characteristic frequency of iteration drops by many orders of magnitude.

This slowing down isn't arbitrary: it is the physical manifestation of increasing redundancy (larger, heavier, more interdependent structures) and increasing stability (longer coherence time).

Astronomical Orbits Frequencies

In terms of the VGF:

- Each stabilisation forms a new coherence timescale — the characteristic time over which the local structure can change while preserving stability.
- In atomic or quantum systems, coherence times are femto- or attoseconds; in planetary systems, years or millennia.
- That slowing is can be seen as a signature of nested stabilisation: redundancy so high that change propagates only very slowly.

Effectively, although "explained" by classical law structures in physics, as due to great mass, and so on, the orbital timescales of stars and planets can also be seen as descendants of earlier coherence frequencies. They can be seen as very slow residual oscillations of a generative process that has condensed through countless stabilisations.

Phenomenon	Frequency / Period

Earth's rotation	$\approx 10^{-5}$ Hz (1 day)
Earth's orbit	$\approx 3 \times 10^{-8}$ Hz (1 year)
Pulsars	$1 - 10^{3}$ Hz (milliseconds to seconds)
Cepheid variables	$10^{-6} - 10^{-4}$ Hz
Binary stars	$10^{-8} - 10^{-5}$ Hz
Spiral galaxy rotation	$10^{-15} - 10^{-17}$ Hz

In due course, we will look at how Hilbert space can be derived from the VGF. The classical laws of physics can be seen as codifications of the stable patterns that emerge after nested stabilisations through the VGF and quantum decoherence have already occurred.

Newton's laws describe the behaviour of stable orbits, but the existence of stable orbits in the first instance arises from decoherence into a classical phase space. In the VGF framework, that stability — the availability of well-defined positions, masses, and momenta — is not just the result of quantum decoherence, but the end result of decoherence cascades.

So orbital and rotational periods are the final, macroscopic echoes of an iterative process that we can suppose began, in terms of our current understanding, at Planck frequency.

Conceptual Synthesis

Scale	Description	Frequency trend	Interpretation in VGF
Quantum	rapid iteration within full coherence	very high	raw generative field
Molecular	local decoherence, vibrational stability	lower	early local completions
Macroscopic	thermodynamic redundancy	lower still	objects & classical behaviour
Planetary	gravitational redundancy	very low	loops of loops—meta-stabilisations
Astronomical	network of stabilised masses	ultra-low	highest redundancy, deepest decoherence

Hence, cosmic slowness is the limit case of stability, the ultimate decoherence as far as matter is concerned: structures so redundant that their "frequency of iteration" in the IIP descriptor tends toward zero.

Review

- Within overall complexity, the nested stabilisations of the VGF descend through frequency domains.
- Each level of redundancy slows the characteristic frequency of iteration: quantum oscillations condense into molecular vibrations, molecular rhythms into macroscopic cycles, and eventually classical frequencies into astronomical orbits.
- Planetary and stellar rotations can therefore be seen as the slowest surviving echoes of the primordial generative field — the last stabilised loops of iteration.
- What physics describes as a "classical law" can be seen as the codified behaviour of those loops once coherence has long since been lost.

Uncertainty and the VGF

The IIP-VGF approach gives us a context for a root understanding of both indeterminacy (uncertainty) in quantum mechanics, and incompatibilities between certain theories as they stand, such as general relativity and quantum theory. It does this by showing where this incompatibility comes from and why it necessarily exists.

Indeterminacy in physics is not chaos and not ignorance. In quantum mechanics, indeterminacy is ontological, not merely epistemic. The uncertainty principle itself is not a statement about what we don't yet know — it's a structural feature of how nature presents measurable properties.

Mathematically:

$$\Delta x \, \Delta p \geq \frac{\hbar}{2}.$$

This is not a measurement error; it is a limit on the coexistence of definability.

It arises because position and momentum are non-commuting operators in Hilbert space. So the "indeterminacy" is a manifestation of non-commutability of relational definitions.

In the VGF Picture

In the IIP-VGF framework, the primitives are relations between iterations — there is no pre-existing substrate of "things" that carry definite values.

So the "indeterminacy" is built in from the beginning, in the sense that: before any stabilisation (decoherence), there is no domain of distinct observables at all — only an open field of possible relations.

When local completions (decoherences) happen, they generate mutually incompatible stabilisations: each completion is a partial closure of infinite iteration, **which inevitably excludes others**.

This is the VGF counterpart to the non-commuting observables of quantum theory:

- x and p are incompatible because they correspond to distinct stabilisation mappings of the same underlying IIP flow.
- Each stabilisation achieves closure in a different "direction" of iteration space — you can't realise both at once.

So, in the VGF sense also, indeterminacy isn't "hidden variables" nor "random chaos". Rather, it's the topological necessity that not all local completions can be simultaneously closed.

Where Determinacy Re-Enters

Classicality appears when redundant stabilisations overlap enough that:

- the incompatible directions of iteration become practically commutative;
- interference terms (off-diagonals) vanish under redundancy;
- a hierarchy of nested completions locks in one network of relations.

This produces stable determinacy — but not by reversing indeterminacy. Rather, determinacy is the limit of redundancy, the state where further iteration no longer changes relative distinctions.

So, the uncertainty principle is preserved:

- At small redundancy (quantum scale): incompatible stabilisations = true indeterminacy.
- At large redundancy (classical scale): effective commutativity = apparent determinacy.

Analogy: Projection of an Infinite Object

We can think of the VGF as an infinite multidimensional pattern of relations. Each "measurement" corresponds to projecting that field into a finite domain.

Different projection directions (*e.g.*, position *vs* momentum) can't capture the same relations coherently because they are, so to speak, different slices through an infinite object.

Thus the indeterminacy is geometric: it is not randomness, but the natural incompatibility of different projections of infinite iteration.

Compatibility With Quantum Mechanics

The VGF doesn't imply "measurements are secretly determined underneath".

Rather, it says that determinacy itself is **the emergent property of decohered iteration**.

At the VGF level:

- There are no hidden variables — because there are no variables at all before stabilisation.
- There's no contradiction with uncertainty — because incompatible stabilisations are the very origin of uncertainty.
- Decoherence is not a deterministic "mechanism" that bypasses quantum indeterminacy; it's the natural process by which indeterminate relational fields self-limit into determinate attractors.

Basic Formulation

Level	Description	Type of indeterminacy
IIP / VGF	Infinite relations without closure	Topological / relational indeterminacy (no defined observables)
Decoherence	Local stabilisation via redundancy	Contextual selection of compatible subrelations
Classical	Stable overlap of redundant stabilisations	Apparent determinacy (effective commutativity)

Determinate classical order emerges when redundant stabilisations overlap enough to make those non-commutativities effectively vanish.

The uncertainty principle, therefore, is not violated by the principle of the VGF, but rather, in the IIP-VGF formulation it is re-derived as a statement about the incompatibility of local completions of infinite iteration.

Indeterminacy as Non-Commutative Completion in the VGF Framework

So in the IIP–VGF picture, indeterminacy is neither ignorance nor hidden variables nor chaos. It arises because different local completions (stabilisations) of the Vast Generative Field (VGF) are mutually non-commuting. Classical determinacy emerges only when redundancy makes those non-commutativities effectively vanish.

From Infinite Iteration to Local Completion

Let $\mathscr{V}$ denote the VGF: an infinite relational field generated by iteration on iteration (IIP). A local completion is a stabilising map

$$\Pi_\alpha : \mathscr{V} \to \mathcal{O}\alpha,$$

that contracts open relational structure into a finite, observable domain $\mathcal{O}\alpha$ (a pointer sector, measurement context, or coarse-grained algebra). Distinct observables (*e.g.* position x and momentum p) correspond to distinct completions Π_x, Π_p.

Key claim. For incompatible observables, the completion maps do not commute:

$$\Pi_x \circ \Pi_p \neq \Pi_p \circ \Pi_x.$$

This non-commutativity is the VGF counterpart of $[\hat{x}, \hat{p}] \neq 0$.

Quantum Correspondence

In standard quantum mechanics (QM), incompatibility is encoded by operator commutators:

$$[A, B] \neq 0 \implies \Delta A \, \Delta B \geq \frac{1}{2} |\langle [A, B] \rangle| .$$

Measurement is modelled by completely positive trace-preserving (CPTP) maps $\mathscr{M}\alpha$ (POVMs or projections

$$\{P_i^{(\alpha)}\}) : \rho \mapsto \mathscr{M}\alpha(\rho) = \sum_i M^{(\alpha)} i \, \rho \, M^{(\alpha)\dagger} i .$$

Incompatibility implies

$$\mathscr{M}\alpha \circ \mathscr{M}\beta \neq \mathscr{M}\beta \circ \mathscr{M}\alpha,$$

generating the same uncertainty constraints.

VGF–QM dictionary:

$$\Pi_\alpha \leftrightarrow \mathscr{M}\alpha, \qquad \Pi\alpha \circ \Pi_\beta \neq \Pi_\beta \circ \Pi_\alpha \leftrightarrow [A, B] \neq 0.$$

Decoherence as Contraction to Redundant Stability

Let $\mathscr{D}_R$ denote the redundancy-indexed decoherence map (environment-induced coarse-graining), with $R \geq 0$ quantifying environmental copying/records:

$$\rho \mapsto \mathscr{D}_R(\rho), \quad \text{with } \mathscr{D}_R : \text{ off-diagonals} \to e^{-\Gamma R} \times \text{off-diagonals} .$$

As $R \to \infty$, coherences in a chosen pointer basis vanish, yielding an effectively classical ensemble.

Effective commutativity under large redundancy.

For two completion channels, define their commutator norm after decoherence:

$$C_{\alpha\beta}(R) := \left\| \mathscr{D}R \circ \mathscr{M}\alpha \circ \mathscr{M}_{\beta}\mathscr{D}R \circ \mathscr{M}\beta \circ \mathscr{M}\alpha \right\|.$$

Then, generically,

$$\lim R \to \infty C_{\alpha\beta}(R) = 0,$$

i.e. redundancy drives effective commutativity (classicality) even though the underlying observables remain non-commuting.

Uncertainty as Incompatibility of Completions

Because different Π_α exclude one another's fine-grained stabilisations, no single completion can preserve full fidelity to all relational directions of $\mathscr{V}$. This yields the uncertainty trade-off:

$$\Delta_\Pi(\alpha)\, \Delta_\Pi(\beta) \gtrsim \frac{1}{2} \left| \langle [A_\alpha, A_\beta] \rangle \right|,$$

where $\Delta_\Pi(\alpha)$ is the dispersion within the Π_α –stabilised sector. Thus, indeterminacy is structural: not ignorance, but the non-commutativity of completion maps.

Emergence of Determinacy

Determinacy is the limit $R \to \infty$ where redundant records make incompatible completions practically commute within the recorded sector:

$$\mathscr{D}R \circ \Pi\alpha \approx \Pi_\alpha \circ \mathscr{D}R, \quad \mathscr{D}R \circ \Pi\alpha \circ \Pi\beta \approx \mathscr{D}R \circ \Pi\beta \circ \Pi_\alpha.$$

Classical laws are then stable relations on the redundancy-dominated manifold.

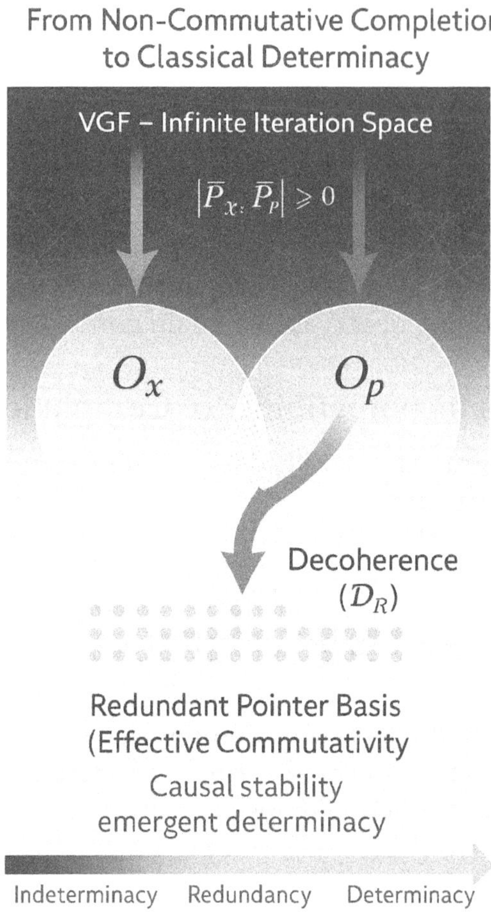

Figure: The VGF derives from infinite iteration space (the IIP), and the very nature of decoherence into stabilisations within the VGF. This is what gives rise to mutually exclusive stabilisations such as position and momentum, and redundant pointer bases.

The IIP-VGF approach gives us a context for a root understanding of uncertainty or indeterminacy in quantum mechanics, by showing where it comes from and why it necessarily exists.

From the VGF to Hilbert Space

Proposition: Karoubi Completion of the VGF Yields the Hilbert-Space Representation

Setup

Let VGF be a relational field in which loops act as idempotent stabilisations $p : X \to X$ with $p \circ p \simeq p$. Let Kar(VGF) denote the Karoubi envelope (idempotent completion) of VGF. Let Hilb be the (idempotent-complete) category of Hilbert spaces and bounded linear maps, where idempotents are orthogonal projectors with images closed subspaces.

Claim

There exists a structure-respecting realisation of stabilised loops as objects: any functor F:VGF→Hilb that sends idempotent loops in VGF to (approximately) idempotent endomaps in Hilb uniquely extends (up to unique isomorphism) along the Karoubi envelope to a functor F̄:Kar(VGF)→Hilb that sends each split idempotent (X, p) to the corresponding projector P and hence to its subspace Im(P). In particular, "emergent objects" = split idempotents in Kar(VGF) are realised concretely as Hilbert subspaces.

Proof Sketch

1) Idempotent completion: By construction, Kar(VGF) freely adds splittings of all idempotents p. Objects are pairs (X, p) with $p^2 = p$; morphisms $f : (X, p) \to (Y, q)$ satisfy $f = qfp$.

2) Extension property: Given F:VGF→Hilb, define $\bar{F}(X, p)$:=Im(F(p))⊆F(X) and $\bar{F}(f)$:=restriction of f to Im(F(p)) using $f = qfp$.

This is functorial and makes $\bar{F}$ the universal extension of F that sends each idempotent to its splitting.

3) Idempotent completeness of Hilb: Every projector P in Hilb splits as F(X)→Im(P)→F(X), so the images of idempotents are bona fide objects (closed subspaces).

Therefore stabilised loops (idempotents) in VGF become objects after Karoubi completion and are realised as subspaces under $\bar{F}$ in Hilb.

Corollary (Physics Gloss)

If VGF encodes the generative relational fabric, then applying the Karoubi envelope is the justified "objectification step," and Hilbert space is its canonical realisation: projectors ↔ split idempotents; subspaces ↔ emergent objects.

Remark

The same argument works in enriched/∞-categorical settings: idempotent completion (*a.k.a.* Cauchy completion) exists under mild hypotheses, and Hilb remains idempotent complete, so the bridge persists at higher categorical levels.

Hilbert Space as the Karoubi Envelope of the Vast Generative Field

VGF (Vast Generative Field)

Primitive = relations (edges).
Loops form redundancies: $p \circ p = p$.
These are idempotent attractors in the raw relational field.
But: in the VGF, these loops don't yet have 'objects' attached — they are just stabilised patterns of relation.

Karoubi Envelope of the VGF

We now construct the Karoubi envelope of this relational field.
Every idempotent loop (X, p) is promoted to a genuine object.
Morphisms are constrained to respect the idempotent structure ($f = q \circ f \circ p$).
Result: we now have a category where objecthood is guaranteed whenever redundancy stabilises.

Hilbert Space as a Karoubi-Complete Setting

In Hilbert space theory:

- Idempotents are orthogonal projectors $P{:}H \rightarrow H$ with $P^2 = P$.
- Every such projector corresponds to a closed subspace of H.
- Those subspaces already exist as genuine objects in the category of Hilbert spaces.

That means: Hilbert space is naturally idempotent complete.
No hidden loops remain unrecognised — every stabilised redundancy has its corresponding subspace.

The Bridge

Hilbert space is the Karoubi envelope of the Vast Generative Field.

- The VGF supplies loops (idempotent redundancies).
- The Karoubi envelope promotes them into objects.
- Hilbert space is the structured arena where those objects appear as subspaces, with the inner product capturing their overlap and orthogonality.

Philosophical Gloss

This perspective reframes Hilbert space not as an arbitrary mathematical invention but as the completion of a deeper generative process.

What we see as 'quantum states' (vectors, subspaces, projectors) are the Karoubi shadows of redundant loops in the VGF.
Objectivity and classicality (in Euclidean space) then arise as further

stages: decohered images of these subspaces, stabilised by redundancy in the environment (Stability–Fidelity Law).

In summary: VGF → Loops (idempotent attractors) → Karoubi envelope (objects from idempotents) → Hilbert space (projectors/ subspaces) → Euclidean space (classical limit).

Flow Diagram

The emergence of Hilbert space, and then Euclidean space, from the Vast Generative Field can be summarised as:

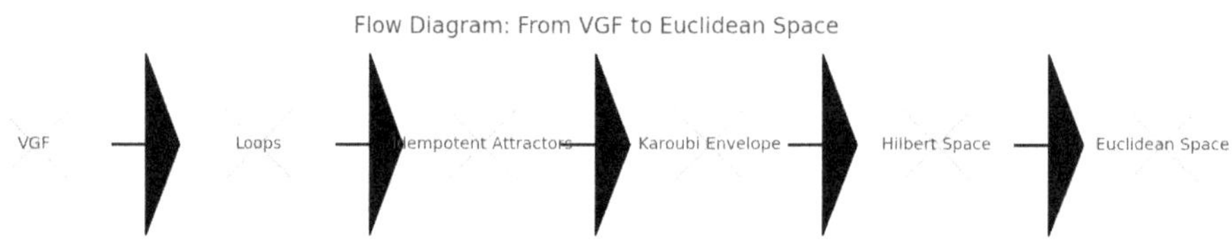

From Hilbert Space Facts Back to VGF Implications

This table summarises how features of Hilbert space constrain or inform our understanding of the Vast Generative Field (VGF). The left column lists established facts about quantum states in Hilbert space; the right column shows the corresponding implications for the ontology of the VGF.

Hilbert Space Fact	VGF Implication
Superposition & linearity	VGF must support a compositional way of combining loops that Karoubi+linearisation turns into vector addition.
Inner product & interference	Loops must carry a U(1)-type phase so overlaps can add or cancel (interference).

Complex scalars (ℂ)	Implies a U(1) phase structure and continuity to support unitary flows (time evolution).
Orthogonality / projectors	VGF must have plentiful idempotent loops that split; these become subspaces.
Born rule / probability	VGF supports a non-Boolean logic of attractors with a measure-like structure on contexts.
Contextuality	No global valuation; VGF organised as local contexts glued sheaf / topos-style.
No-cloning / no-broadcasting	Copying possible only for commutative substructures (pointer attractors).
Tensor product & entanglement	VGF must compose via non-Cartesian monoidal structure allowing non-factorisable loops.
Unitary evolution	VGF should admit a dagger / involution capturing reversibility of coherent transformations.
Spectral theorem	VGF supports maximal commuting families of split idempotents (contexts with spectral decomposition).
Mixed states & CP maps	VGF must admit a probabilistic / operational layer; decoherence = CP maps.
Superselection sectors	VGF likely graded by a group, fencing off loop families (non-interfering sectors).
Symmetry (Wigner)	VGF admits automorphisms that become unitaries / antiunitaries preserving overlap.

The Five Distinctions of Hilbert Space

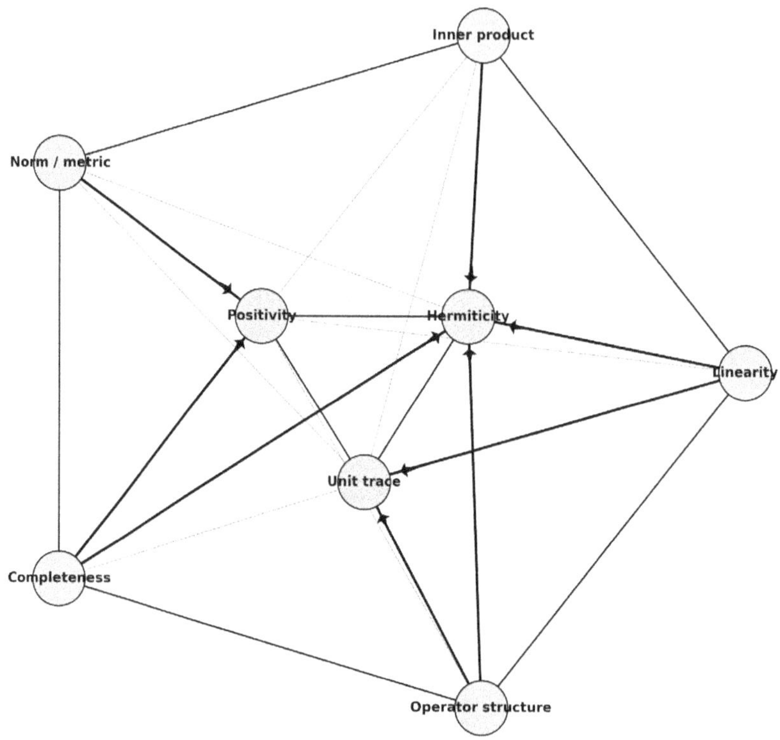

This diagram shows the five Hilbert space discriminations (outer nodes) and the three defining density matrix properties (inner nodes) as a fully connected directed network. The outer pentagon and inner triangle are joined to emphasise that neither set of nodes exists in isolation. Directed arrows indicate how discriminations contribute to stabilising properties.

Thick black arrows represent strong/primary contributions (e.g., Inner product → Hermiticity, Norm/metric → Positivity, Linearity/Operator structure → Unit trace). Dashed grey arrows represent weaker or background connections, reinforcing the idea that the whole network acts as a discriminative web stabilising the density matrix.

In the IIP-VGF framework this is a web of stabilisations or closures in the decoherence of the VGF.

From IIPs to Measurement Outcomes

Infinite Iteration Processes (IIPs)

The question of the evolution and nature of neural intelligence and its connection with decoherence, can be framed in terms of the infinite iteration principle and decoherence from its consequent field of relations - the VGF.

We begin with Infinite Iteration Processes (IIPs) as primitives. An IIP is the abstract principle of iteration signified as ⊙ and defined only by its capacity infinite self-iteration. It embodies the idea of pure recurrence, prior to the emergence of physical time.

The Superiterator

The original IIP ⊙ is an iteration-of-iterations. Instead of generating outputs directly, ⊙ generates entire IIPs. Thus, the infinity of distinct IIPs is not a static set but the orbit of ⊙, from which the VGF and its consequences are derived.

Formal sketch:

$$\odot \ : \ N \rightarrow \text{IIPs, with } \odot (n+1) = \mathscr{T}(\odot (n)),$$

where $\mathscr{T}$ is a meta-iteration operator.

Hilbert Space as a Derivative

From ⊙, we can construct correlations between its generated IIPs. These correlations induce a Hilbert space $H \odot$ via the kernel:

$$K(i,j) = lim(N \rightarrow \infty)(1/N) \odot (n = 1toN) f_i(n) f_j * (n)$$

Hilbert space therefore appears as a secondary structure arising from ⊙, encoding the relational fabric of the generated IIPs.

Decoherence and Pointer States

In quantum Darwinism decoherence acts as a redundancy filter. From the vast potential correlations of Hilbert space, it stabilizes a subset of robust, redundant modes: the pointer states. These are the survivors of iteration in the physical domain, forming the objective "menu" of possible classical measurement outcomes.

The fact that an actual classical measurement outcome in a laboratory experiment remains indeterminate, despite the current understanding of decoherence, persists as **the measurement problem**.

The Measurement Problem Reframed

Neural intelligence in humans arising through brain function must itself be considered as a decoherence-stabilized structure. It can also be proposed as arising through nested redundancies in neural networks, memory, and learning.

Physics alone provides the pointer basis, but not the specific measurement outcome. "Traditional" approaches have been been asking: 'Which outcome is chosen by nature?'

In quantum Darwinism the objectivity of measurables in classical science is a consequence of the redundancy of pointer states, and consensus, rather than being simply indicative of separateness from the observer.

In the IIP framework, quantum decoherence and the evolution of the species and consequently neural intelligence, are all embraced together and describable through the same principle of generalised decoherence of the VGF.

Synthesis

In the IIP framework, because the IIP principle ⊙ is the primitive itself, and because its consequent VGF leads to attractors and then decoherence from the VGF in both quantum and classical domains, then the principle embraces both:

- The arising of objective pointer states (physics).
- The evolution of discriminatory intelligence (cognition).

Measurement outcome is then the **joint stabilization of these two**, not as collapse imposed from outside, but as **co-emergence within the same iterative root**.

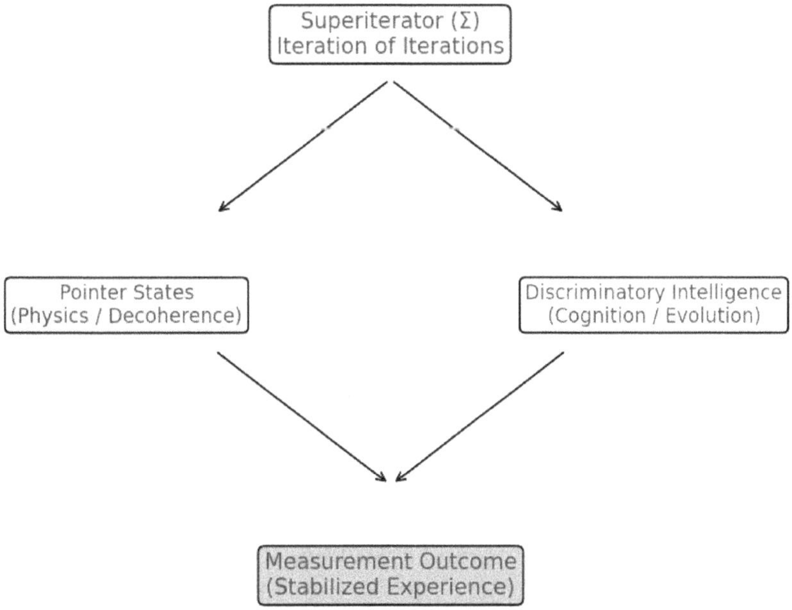

Figure: The IIP branching into Physics (pointer states) and evolutionary intelligence, converging in measurement outcome.

Quantum Darwinism

Nested and Networked Completions: From the VGF to Hilbert Space and Quantum Darwinism

Arising from the principle of infinite iteration, ⊙, the Vast Generative Field (VGF) represents infinite generativity — iteration without end, the principle of the IIP itself being total **coherence without distinction**.

Within this field, local stabilisations emerge where iteration becomes self-consistent, forming idempotent attractors or 'local completions.' Each local completion retains residual coherence and can itself undergo further stabilisations, producing a hierarchy of completions within completions.

At the same time, these stabilisations interact, forming networks of mutual reinforcement. The combination of **nesting** and **networking** gives rise to the structures we recognise as the mathematical and physical frameworks of Hilbert space and Quantum Darwinism.

From Generativity to Local Completions

At the highest level, the VGF has no boundaries or partitions. It is a field of infinite coherence—relations iterating upon themselves without closure. Local completions occur when self-referential iteration achieves stability: an idempotent state in which further iteration reproduces itself ($x^2 = x$). These completions constitute the first local decoherences, stabilised forms within an otherwise unbounded field.

Nested and Networked Structure

Each completion retains partial coherence, allowing new distinctions to form within it. This creates a nested hierarchy of stabilisations: completions within completions, each layer producing further idempotent substructures. Simultaneously, these completions interact, exchanging information and stabilising one another through networks of causal relations. Together, nesting and networking generate the conditions for structured coherence—the domain that can be modelled by Hilbert space.

Hilbert Space and Quantum Darwinism as Emergent Layers

When networks of completions reach a sufficient degree of relational coherence, they can be represented as vectors and operators within Hilbert space. This is the formal crystallisation of coherence relations between locally stable systems. Within this domain, further partitioning — system and environment — leads to decoherence and redundancy. The emergent pointer basis of Quantum Darwinism represents the outermost layer of this hierarchy: redundant, stable records of nested and networked completions.

Stability–Fidelity Hierarchy

Level	Description	Dominant Property
VGF	Infinite generativity; no partitions or limits	Maximum fidelity, no stability
Local completions	First stabilisations (idempotent loops, proto-objects)	Local stability begins
Nested completions	Recursive decoherences within completions	Hierarchical stability
Networked completions	Interacting stabilisations (causal networks)	Redundancy and coherence exchange
Hilbert space	Formal representation of coherent relations	Balanced stability–fidelity regime
Quantum Darwinism	Redundant broadcasting of stable pointer states	Maximum stability, minimal fidelity

Nested and Networked Completions

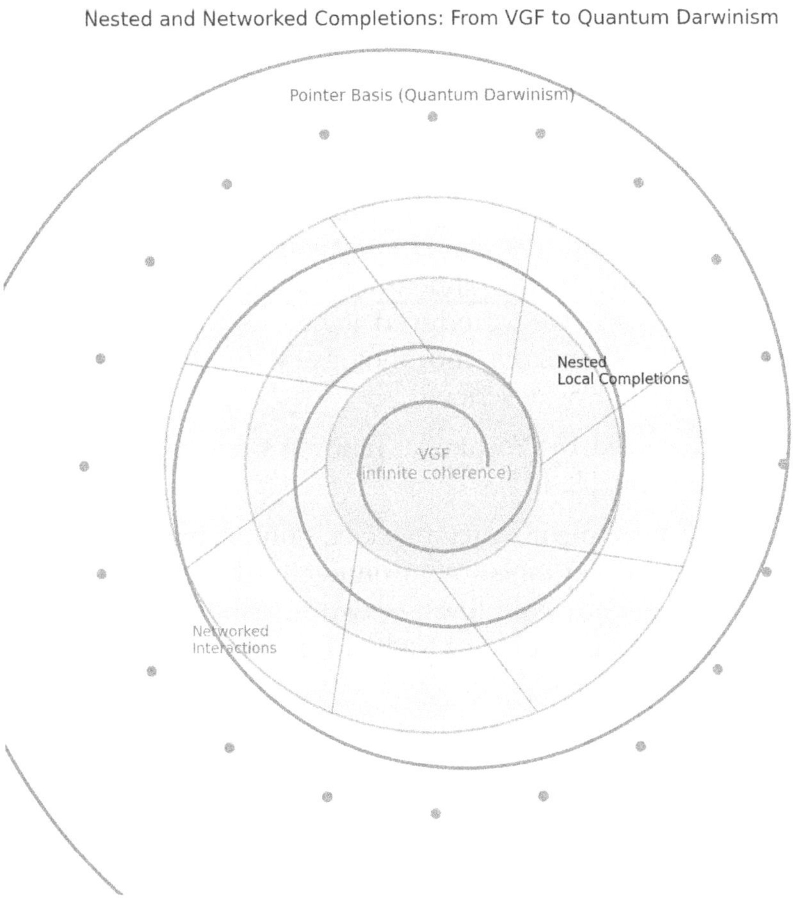

Figure: Conceptual schematic of the descent from the Vast Generative Field (VGF) to Hilbert space and Quantum Darwinism. The inner spiral represents infinite generativity (VGF). Grey shells denote nested local completions — bounded stabilisations within the field. Radiating lines linking the grey shells indicate networked interactions among completions. The dotted outer layer represents the pointer basis of Quantum Darwinism: These are the redundantly stabilised, classical-like states formed at the terminus of the Stability–Fidelity descent.

The HEOM

The **Hierarchical Equations of Motion** (HEOM) formalism is one of the most powerful and general approaches for describing the dynamics of open quantum systems — systems that interact with an external environment (bath) and thus exhibit decoherence and dissipation. The reason we introduced it here is that as we will be seeing the HEOM shares its morphology with decoherence and stabilisation in the VGF as described.

It was first introduced by Yoshitaka Tanimura and Ryogo Kubo in 1989 (J. Phys. Soc. Jpn. 58, 101).

Their innovation was to construct an exact, non-perturbative method that could treat non-Markovian environments (those with memory) and strong system–bath coupling — two situations where earlier master equations (like Lindblad or Redfield) failed.

Key Sources and Developments

Y. Tanimura & R. Kubo (1989): Original derivation of the hierarchy (J. Phys. Soc. Jpn. 58, 101).

Y. Tanimura (1990, 1991, 2014): Subsequent refinements and efficient numerical formulations (Phys. Rev. A 41, 6676; J. Chem. Phys. 141, 044114).

A. Ishizaki & G. R. Fleming (2009): Application to excitonic energy transfer; evidence for long-lived coherence in biological systems (PNAS 106, 17255).

M. Yan, Y. Tanimura et al. (2018–2024): Generalised and tensor-network extensions of HEOM for large systems and low temperatures.

Tanimura, Y. & Kubo, R. (1989). Time evolution of a quantum system in contact with a nearly Gaussian-Markoffian noise bath. J. Phys. Soc. Jpn. 58, 101–114.

Tanimura, Y. (2014). Real-time and imaginary-time quantum hierarchal Fokker–Planck equations. J. Chem. Phys. 141, 044114.

Ishizaki, A. & Fleming, G. R. (2009). Theoretical examination of quantum coherence in photosynthetic energy transfer. Proc. Natl. Acad. Sci. USA 106, 17255–17260.

Diósi, L. & Strunz, W. T. (1997). The non-Markovian stochastic Schrödinger equation for open systems. Phys. Lett. A 235, 569–573.

Tanimura, Y. (2020). Numerically "exact" approach to open quantum dynamics. J. Chem. Phys. 153, 020901.

Basic Idea

In the Hierarchical Equations of Motion (HEOM) framework, 'motion' does not describe the displacement of a particle in Euclidean or relativistic spacetime. Instead, it refers to the continuous evolution of the system's density operator within an abstract space of states. The differential $d\rho/dt$ expresses change with respect to time, but this time is **the parameter of generative unfolding, not necessarily physical or geometric time**.

Evolution in Liouville Space

The HEOM formalism operates in Liouville space—the vector space of operators. Each point in this space represents an entire quantum state ($\rho^{(0)}$, $\rho^{(1)}$, $\rho^{(2)}$, …), including auxiliary density operators that encode environmental memory. Motion in this space means a continuous reconfiguration of the informational and coherence structure of the system. The hierarchy of auxiliary operators forms a network of coupled layers, through which coherence flows and decays.

Nature of Motion

In HEOM, motion is the transformation of state amplitudes across the hierarchy, driven by both the Hamiltonian generator and the environment couplings. It is not spatial translation but the evolution of correlations and coherences across nested operator spaces. The quantity that evolves is informational rather than positional—an abstract form of motion within the manifold of states.

VGF Interpretation

From the perspective of the Vast Generative Field (VGF), the HEOM hierarchy is a dynamic structure of local completions linked through generative iteration. Each layer ($\rho^{(0)}$, $\rho^{(1)}$, $\rho^{(2)}$, ...) represents a local stabilisation, while the couplings between them express continuing generativity. Time in this context is not external but intrinsic: it is the index of iteration itself—the ordering of coherence transformations within the generative field.

Thus, the HEOM embodies the same principle as the VGF: motion is not traversal of space, but transformation through nested relations and evolving coherence. The system's evolution is a morphogenesis of informational form.

Motion in Operator (Liouville) Space

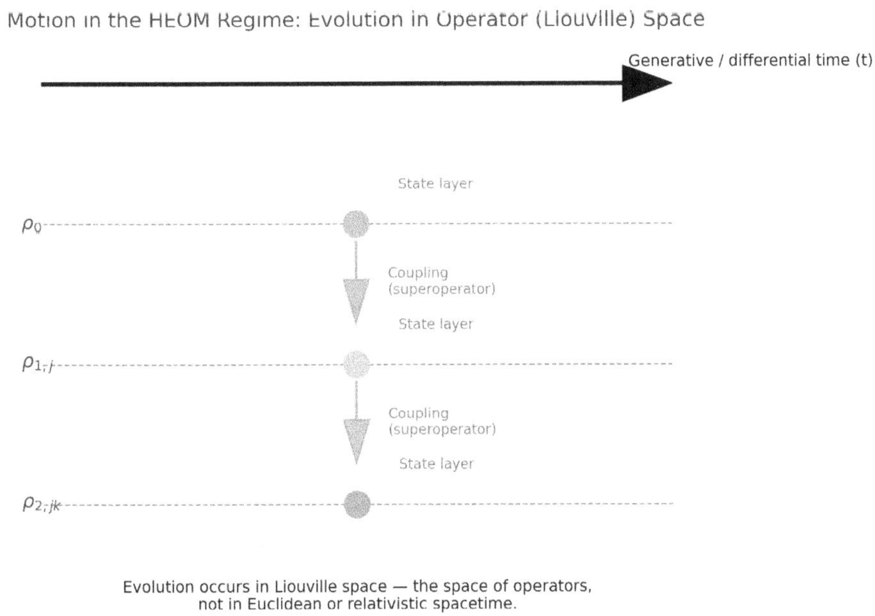

Figure: Motion in the HEOM regime occurs through Liouville space—the space of operators—rather than through Euclidean or relativistic space. Each state layer (ϱ_0, ϱ_1, ϱ_2, …) evolves under time as a generative parameter, with vertical couplings representing superoperator interactions between memory layers. The horizontal axis denotes generative or differential time t, marking iteration within the hierarchy.

In a typical open-system setup, the total Hamiltonian is written as:

$$H_{\text{tot}} = H_S + H_B + H_{SB},$$

where H_S is the system, H_B the bath (environment), and H_{SB} their interaction.

Tracing out the environment to obtain a reduced density operator $\rho_S(t)$ generally yields an integro-differential equation with memory kernels that are very difficult to handle directly.

Tanimura and Kubo's insight for the HEOM was to expand the bath correlation function — typically a sum of exponentials — and to transform the resulting nonlocal memory terms into a hierarchy of coupled ordinary differential equations for a set of Auxiliary Density Operators (ADOs):

$$\rho^{(0)}(t) = \rho_S(t), \quad \rho^{(1)}j(t), \quad \rho^{(2)}jk(t), \quad \ldots$$

Each higher-order ADO stores information about progressively deeper levels of system–environment correlation.

The Hierarchical Structure

The hierarchical equations can be given as:

$$\dot{\rho}^{(n)}(t) = -\left(i\mathscr{L}_S + \sum_j n_j\gamma_j\right)\rho^{(n)}(t) \\ -i\sum_j [V, \rho_j^{(n+1)}(t)] \\ -i\sum_j n_j\left(c_jV\rho_j^{(n-1)}(t) - c_j^*\rho_j^{(n-1)}(t)V\right).$$

Annotations (term by term)

Symbol	Meaning	Physical / Conceptual Interpretation
$\rho^{(n)}(t)$	n^{th}-tier Auxiliary Density Operator (ADO)	Describes the system's state entangled with the environment to correlation order n . $\rho^{(0)}$ is the system's observable state; higher tiers encode **memory and feedback**.
$\mathscr{L}_S$	System Liouvillian superoperator: $\mathscr{L}_S X = [H_S, X]/\hbar$	Generates the system's *unitary evolution*; internal dynamics without environment.
γ_j	Damping constants associated with bath modes or correlation functions	Define how quickly memory in the environment decays — how "Markovian" or "non-Markovian" the dynamics are.

V	System operator that couples to the bath (interaction operator)	Specifies how the system and environment interact — e.g. position, dipole moment, spin operator.
c_j	Expansion coefficients in the decomposition of the bath correlation function	Complex weights that set the amplitude and phase of memory feedback; derived from the bath spectral density.
n_j	Integer index tracking how many times the system has interacted with bath mode j	Counts the order of correlation between system and that bath component.
$[V, \rho_j^{(n+1)}]$	Coupling up the hierarchy	Describes how correlations with the environment add new coherence to the system (forward coupling).
$n_j(c_j V\rho_j^{(n-1)} - c_j^* \rho_j^{(n-1)} V)$	Coupling down the hierarchy	Describes how existing correlations decay, feeding coherence back to the system as stabilisation.

Each level n of the hierarchy couples upward to finer environmental correlations (n+1) and downward to coarser ones (n-1).

In practice the hierarchy is truncated at some finite level N, which yields numerically convergent dynamics for systems such as quantum dots, molecular excitons, or spin-boson models.

What HEOM Achieves

The HEOM approach has become a benchmark for exact open-quantum-system dynamics, widely applied in:

- Condensed-phase spectroscopy — capturing vibrational coherence and energy transfer in photosynthetic complexes (Ishizaki & Fleming, PNAS 106, 17255 (2009));
- Quantum transport — treating electron flow through molecular junctions and quantum dots;

- Quantum information — modelling decoherence of qubits in structured environments;
- Chemical dynamics — describing solvated reactions and proton transfer.

Because it explicitly retains environmental memory, HEOM bridges the gap between unitary quantum mechanics and irreversible classical behaviour without relying on phenomenological damping.

Morphological Significance

Conceptually, the hierarchy of ADOs in HEOM forms a temporal nesting: each tier represents a slower stabilisation of coherence than the one beneath it.

At the same time, the coupling operators linking different ADOs constitute a network of correlations.

Thus, the mathematical structure of HEOM simultaneously expresses:

- Nesting: vertical layering of timescales (short-term quantum → long-term classical memory),
- Networking: horizontal coupling of modes and operators (interaction pathways).

This duality makes HEOM particularly resonant with the VGF framework, in which coherence condenses into structure through nested and networked stabilisations.

Review

The Hierarchical Equations of Motion transform the problem of tracing out an environment with memory into a deterministic cascade of coupled density operators.

Each layer of the hierarchy corresponds to a deeper level of stabilisation of coherence — a nested memory — while the interlayer couplings define a network of dynamical relations.

In this way, the HEOM formalism mathematically embodies the same morphology that the Vast Generative Field describes conceptually: coherence condensing into structure through nested and networked iteration.

Interpretive Summary - VGF Similarity

The HEOM hierarchy represents a **nested flow** of information: coherence propagates upward into memory layers, decays through exponential damping, and feeds back into the system as effective dissipation. In quantum systems, ϱ_n are operators carrying phase coherence; in classical systems, they are probability distributions; in the VGF framework, they correspond to local completions within an iterative field of generative relations.

The full hierarchy embodies the principle of recursive stabilisation: reversible evolution ($\mathcal{L}$) coexists with irreversible memory loss (γ_j), yielding emergent stability and effective laws at the lowest level (ρ_0).

The HEOM and the IIP-VGF

The HEOM and the IIP-VGF framework are related but different. The hierarchical equations of motion describe the iteration of states within a fixed generator.

In contrast, in the VGF

$$\mathscr{K}_{n+1} = g(\mathscr{K}_n)$$

describes the iteration of the generator itself — *i.e.*, iteration at the level of the law or coupling.

The HEOM: Iteration of State Within a Fixed Generator

The Hierarchical Equations of Motion (HEOM) are dynamical equations for a hierarchy of density operators $\rho^{(n)}(t)$:

$$\dot{\rho}^{(n)} = -\left(i\mathscr{L}_S + \sum_j n_j\gamma_j\right)\rho^{(n)}$$

$$i\sum_j [A, \rho_j^{(n+1)}]$$

$$i\sum_j n_j(c_j A \rho_j^{(n-1)} - c_j^* \rho_j^{(n-1)} A).$$

They are linear differential equations with a fixed operator structure (the Liouvillian $\mathscr{L}_S$, damping rates γ_j, coupling operators A).

Here:

- What evolves is the state ($\rho^{(n)}(t)$);
- The generator ($\mathscr{L}_S, \gamma_j, c_j$) is fixed.

So the HEOM describe how coherence decays and memory propagates given a stable rule of interaction.

In contrast, The $\mathscr{K}_{n+1} = g(\mathscr{K}_n)$ recursion of the VGF is iteration of the generator itself

When we iterate $\mathscr{K}$ upon itself,

$$\mathscr{K}_{n+1} = g(\mathscr{K}_n),$$

and then we are describing how the rule that defines coupling (the generator of stability–fidelity transformation) evolves through its own action.

This is a recursion of generators, not of states.

So:

Hierarchy	Variable being iterated	What it represents
HEOM	state (ϱ)	how coherence evolves under fixed
K-recursion	law ($\mathcal{K}$)	how the law itself evolves and

The HEOM are first-order dynamics within a given framework.

The K-recursion is meta-dynamics of the framework itself — a higher-order recursion.

The Relationship: Meta-Iteration and Renormalisation

In ordinary physical terms, these two levels are related by renormalisation or self-consistency:

- HEOM evolve a system under fixed coupling parameters (γ_j, c_j).
- But in reality, those parameters depend on the system's history — on how coherence and correlation have evolved so far.

So, over long timescales, the effective $\gamma_j$and c_j change.

That is exactly the situation that $\mathscr{K}_{n+1} = g(\mathscr{K}_n)$ captures.

In other words:

The iteration of $\mathscr{K}$ represents **the slow evolution of the generator parameters that underlie the fast HEOM dynamics**.

This relationship is precisely what is meant by **multi-timescale nesting** in open systems and complex adaptive systems. We shall be revisiting this principle when we come to discuss classical evolution.

Nested-Timescale Structure

We can think of this kind of iteration as two interlocked layers of recursion:

Fast layer (state iteration):

$$\dot{\rho} = \text{HEOM}[\rho \mid \mathcal{K} \text{ fixed}]$$

Slow layer (generator iteration):

$$\mathscr{K}_{n+1} = g(\mathscr{K}_n \mid \langle \rho \rangle \text{ feedback})$$

- The fast layer (HEOM) describes quantum or classical evolution with memory.
- The slow layer (K-recursion) describes how the memory kernels themselves adapt through feedback from the evolving ρ.

Mathematically this is akin to two coupled iterative maps — one acting in state space, the other in operator space.

Physically it's the difference between:

- The micro-evolution of dynamics (decoherence, energy flow); and
- The macro-evolution of laws (emergent stability, adaptive learning, renormalisation).

HEOM as a "Frozen" Case of the *K*-recursion

If $\mathscr{K}$ converges (or oscillates stably) under self-iteration, it becomes effectively constant:

$$\mathscr{K}n + 1 = \mathscr{K}n = \mathscr{K}^* .$$

Then HEOM operate under this fixed $\mathscr{K}^*$, describing stable decoherence dynamics.

So standard HEOM are a **limiting case of the higher-order recursion** — they are a "frozen law" phase of the more general iterative process.

The VGF–IIP Interpretation

In VGF terms:

- HEOM correspond to **iteration within a local completion** (how a structure behaves under its own stabilisation rule);
- The $\mathscr{K}$ recursion corresponds to **iteration between completions** (how stabilisation rules themselves iterate and form new attractors).

Thus:

The HEOM are what the VGF looks like inside a stabilised domain;

The $\mathscr{K}$ recursion is what the VGF looks like when that stabilised domain itself becomes part of a higher stabilisation.

Or, said another way:

- HEOM = nested **dynamics within a completion** (intra-level iteration).
- $\mathscr{K}$ recursion = **network evolution across completions** (inter-level iteration).

Comparison

Aspect	HEOM	$\mathcal{K}$-recursion
Domain	State-space (ρ)	Operator-space ($\mathcal{K}$)
Iteration type	Linear (differential)	Nonlinear (map or functional recursion)
Fixed or evolving law?	Fixed	Self-evolving
Timescale	Fast, microscopic	Slow, meta-dynamic
Physical meaning	Decoherence and memory	Renormalisation and emergence
VGF level	Inside a local completion	Evolution between completions
Limiting case	$\mathscr{K}$ = const.	$g(\mathscr{K}) = \mathscr{K}$ (steady attractor)

In short, the HEOM describe the internal, fine-grained iteration of states within a stable generative law; the $\mathscr{K}$-recursion describes the slow iteration of that law itself — the emergence of new stabilisation rules.

Generalised Dynamics

The relationship between them is one of **nesting and feedback**: fast decoherence dynamics feed back into the slow evolution of the coupling structure, producing the self-similar hierarchy that the VGF embodies:

$$\mathscr{K}_{n+1} = \mathscr{F}(\mathscr{K}_n) = \mathscr{K}n^2 \quad \text{(or more generally)} \quad \mathscr{K}n + 1 = g(\mathscr{K}_n)$$

$$\mathscr{K}_{n+1} = g(\mathscr{K}_n)$$

Interpreting g in the Stability–Fidelity / VGF Context

- $\mathscr{K}_n$ represents the current local balance of stability (Γ) and fidelity (Ω) across the web of relations.

- g represents the feedback of those balances on themselves — *i.e.*, the way stabilisation at one level modifies the conditions for stabilisation at the next.

Formally, this is what is model in the general VGF equation:

$$g(\mathcal{K}) = \alpha\mathcal{K}^2 + \beta\mathcal{K} + \gamma I,$$

where the coefficients (α, β, γ) reflect redundancy / fidelity trade-offs.

Superficially this looks trivial, but in operator algebra, this kind of quadratic recursion generates extremely rich behaviour (analogous to the Mandelbrot map).

So g is not arbitrary: it encodes the recursion law of stabilisation.

We could equally well write it as $\mathcal{F}[\mathcal{K}]$, to make explicit that it is a functional rather than a mere function.

Meaning

The **interpretation** is guided by what kind of object $\mathcal{K}$ is:

If $\mathcal{K}$ is...	Then $g(\mathcal{K})$ means...
A matrix / operator	Apply some algebraic rule to its entries (square, trace, commutator, etc.)
A tensor of couplings	Compute the renormalised tensor from interactions among entries
A stability–fidelity balance	Compute next-level balance from recursive application of same trade-off rule

Thus, writing $\mathcal{K}_{n+1} = g(\mathcal{K}_n)$ is shorthand for the recursion of the coupling law itself.

The meaning of g must always be specified by the context — but the notation is the conventional way to express "an iterative functional update".

Conceptual Summary

Aspect	HEOM	$\mathscr{K}_{n+1} = g(\mathscr{K}_n)$
Object updated	state (ρ)	law ($\mathcal{K}$)
g or operator form	fixed linear Liouvillian	nonlinear functional (renormalisation)
Iteration domain	Hilbert/Liouville space	operator space
Goal	track system	evolve the generator itself
Notation	differential recursion	functional recursion

Review

The expression $\mathscr{K}_{n+1} = g(\mathscr{K}_n)$ is a standard way to denote a functional iteration — a recursion where the next operator is obtained by applying a rule g to the current one.

In the present context, g represents the feedback law through which the generator of stability and fidelity modifies itself as stabilisations accumulate.

The form is analogous to renormalisation flows or self-consistent field updates: it expresses the meta-level recursion that acts on the space of generators themselves, not just on states within a generator.

HEOM - Beyond Quantum Applications

The Universal Hierarchical Dynamics of Memory

Introduction

The Hierarchical Equations of Motion (HEOM) are commonly introduced as a tool in quantum mechanics for describing open-system dynamics with non-Markovian environments. Yet the HEOM structure itself—the hierarchy of coupled equations representing nested memory—is not intrinsically quantum. It is a general mathematical framework for modelling systems with feedback, history, and recursion. This universality means that while the HEOM originated in quantum theory, it can equally represent classical stochastic systems or even abstract generative feedback architectures such as those described in the Vast Generative Field (VGF) framework.

The Structure of HEOM

The general HEOM takes the form of a coupled hierarchy:

$$\dot{\rho}^{(n)}(t) = -\left(i\mathscr{L}_S + \sum_j n_j\gamma_j\right)\rho^{(n)}(t) - i\sum_j [A, \rho_j^{(n+1)}(t)] - i\sum_j n_j\left(c_j A\rho_j^{(n-1)}(t) - c_j^* \rho_j^{(n-1)}(t)A\right).$$

Here $\rho^{(0)}(t)$ represents the system's reduced density operator, while higher-tier terms $\rho^{(n>0)}(t)$ denote Auxiliary Density Operators (ADOs) that encode the system's correlations with the environment.

Each tier retains a progressively shorter memory timescale, corresponding to increasingly indirect correlations between system and bath.

Together, the hierarchy models the feedback and decay of correlations in an open system — capturing how coherence is transferred, stored, and ultimately lost.

Quantum–Mechanical Formulation

In the quantum domain, the HEOM are formulated in Liouville space, where density operators act on the system's Hilbert space.

The Liouvillian superoperator $\mathscr{L}_S = [H_S, \cdot\,]/\hbar$ generates unitary evolution under the system Hamiltonian H_S, while environmental influence enters through the commutator terms $[A, \cdot\,]$.

Each auxiliary operator $\rho_j^{(n)}(t)$ represents a specific order of system–bath entanglement, weighted by the bath correlation coefficients c_j and damping constants γ_j.

The hierarchy therefore tracks the flow of coherence through nested entanglement channels, maintaining phase information over multiple memory depths.

When higher tiers decay ($\rho^{(n>0)} \to 0$), coherence is irreversibly transferred to the environment, and the base tier $\rho^{(0)}$ converges toward a decohered steady state.

Classical Limit

In the classical limit, commutators are replaced by Poisson brackets $[A, \cdot\,] \to i\hbar\{A, \cdot\,\}_{PB}$, and the density operators become phase-space probability distributions rather than operators.

The resulting hierarchy reduces to a generalised Fokker–Planck or non-Markovian Langevin system, in which the ADOs correspond to colored-noise memory kernels.

This form preserves the same recursive structure as the quantum case: memory and dissipation remain coupled through a nested feedback chain.

Hence, the HEOM formalism **unifies quantum and classical open-system dynamics** under a single iterative correlation hierarchy —

quantum coherence and classical correlation being two limits of the same underlying recursion.

VGF and Generative Interpretation

In the Vast Generative Field framework, the HEOM morphology exemplifies the iterative logic of stability emerging from recursion. Each $\rho^{(n)}$ corresponds to a local completion within a nested feedback process, in which coherence (or informational correlation) flows outward and becomes stabilised as redundancy. Whether the underlying algebra is quantum, classical, or purely generative, the same pattern appears: **iteration, feedback, and stabilisation across scales**.

HEOM is thus a mathematical archetype of generative recursion itself.

Comparative Overview

Domain	State Variable ($\rho^{(n)}$)	Underlying Algebra	Interpretation of Memory	Typical Example
Quantum HEOM	Density operators on Hilbert space	Commutators [,]	Coherence memory via entanglement	Spin–boson model, exciton transport
Classical HEOM	Probability functions on phase space	Poisson brackets { , }	Trajectory memory via correlation kernels	Generalized Langevin dynamics
VGF Morphology	Iterative feedback objects	Abstract relational algebra	Nested completions and stabilisations	Stability–Fidelity recursion

Morphological Synthesis

Across all interpretations, HEOM expresses the same deeper morphology: coherence, memory, and recursion form a nested sequence of stabilising loops. Quantum mechanics manifests this as coherence feedback between system and bath; classical dynamics as temporal correlations in stochastic trajectories; and the VGF as iteration of generative relations yielding local completions. The distinction lies only in the algebraic layer, not in the structure itself.

HEOM Morphology: Same Recursive Structure, Different Algebraic Realisations

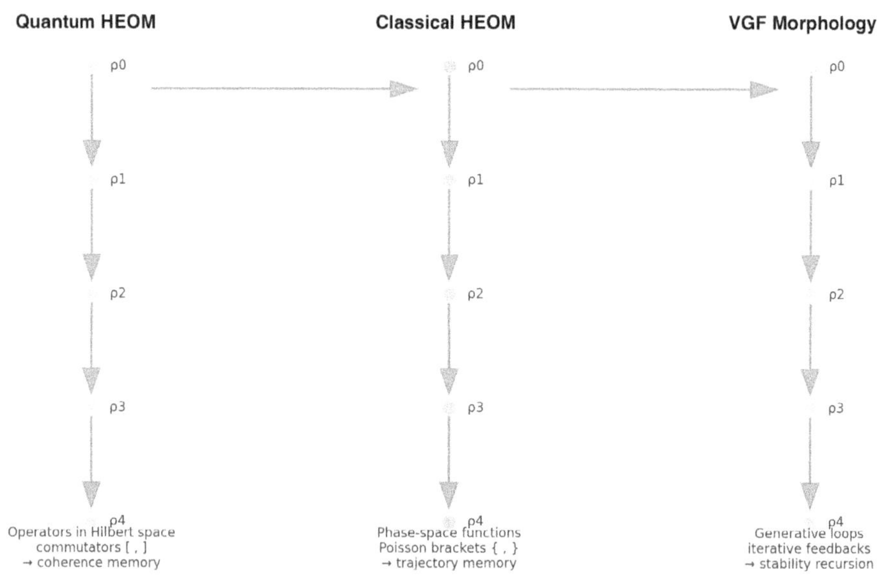

Figure: Three instantiations of the same recursive HEOM structure. **Left**: Quantum HEOM with operator hierarchy in Hilbert space. **Centre**: Classical HEOM with distribution hierarchy in phase space. **Right**: VGF morphology as nested generative feedbacks producing stability. Arrows between columns indicate the formal universality of the recursion—different algebraic realisations of the same generative pattern.

Universality of the HEOM: From Laboratory Decoherence to Natural Stabilisation

The Hierarchical Equations of Motion (HEOM) provide an exact, non-perturbative formalism for describing the evolution of quantum systems interacting with their environments. Originally developed for condensed-phase molecular dynamics, the same formalism applies universally to any open quantum system—laboratory or natural—because it encodes the fundamental process by which coherence is redistributed, stabilised, and ultimately lost. In this sense, HEOM represents not merely a computational scheme but a general morphology of stabilisation in nature.

Laboratory *vs.* Natural Decoherence

Domain	Environment	Timescales	Typical Systems
Laboratory	Controlled reservoirs (phonons, optical cavities, cryogenic baths)	picoseconds–seconds	Molecules, qubits, excitons, superconducting circuits
Natural	Uncontrolled cosmic, biological, or material fields	femtoseconds–cosmic ages	Radiation fields, molecular media, planetary systems, early-universe decoherence

Both follow the same underlying principle:

$$H_{tot} = H_S + H_B + H_S B,$$

where the system Hamiltonian H_S couples through $H_S B$ to an environment H_B whose memory and feedback are represented in the HEOM hierarchy.

Dynamic Boundary Emergence

In laboratory setups, the boundary between system and environment is imposed by experimental design. In natural contexts, it is **emergent** —local subsystems continually act as each other's environments. The HEOM hierarchy automatically adapts to this situation: higher-order Auxiliary Density Operators (ADOs) represent deeper environmental memory, and truncating them defines the effective boundary between coherence and redundancy.

VGF Interpretation

In the Vast Generative Field (VGF) perspective, the HEOM represents a finite mathematical cross-section through an infinite generative process:

HEOM feature	VGF interpretation
System–environment coupling	Local partition of generative field into temporary subloops
Higher-order ADOs	Nested completions carrying environmental memory
Convergence to steady state	Local stabilisation: coherence condensed into redundancy
Universality	Decoherence as a fundamental law of completion, not an artefact of observation

Hence, laboratory and natural decoherence are two realisations of the same generative principle: stabilisation through nested iteration.

Review

Coherence is embodied in relations of relations. Networks embody structure. The HEOM formalism describes the universal dynamics by which **coherence** becomes **structure**. Whether in an optical cavity, a biological molecule, or a cosmic field, the same hierarchy governs the redistribution of **coherence** into **redundant, stable configurations**.

In the VGF interpretation, this is the mathematical expression of **nature's iterative stabilisation**—the continual emergence of the classical world.

Time-Nonlocal Master Equations and the Hierarchy of Motion: Memory, Nesting, and Networking

Actual environments in nature often retain memory, invalidating the Markov approximation underlying the Lindblad equation.

Two complementary approaches generalise Markovian dynamics: (i) time-nonlocal master equations, which introduce an explicit memory kernel, and (ii) Hierarchical Equations of Motion (HEOM), which recast the nonlocal memory into a coupled set of time-local equations by introducing auxiliary density operators (ADOs).

Both approaches make explicit the nested and networked morphology that is only implicit in the Lindblad form.

Time-Nonlocal Master Equations (Nakajima–Zwanzig)

Starting from the Liouville–von Neumann equation for the total system–environment state and projecting onto the system subspace yields an integro-differential equation for the reduced state $\rho_{S(t)}$:

$$\frac{d\rho_{S(t)}}{dt} = -\left(\frac{i}{\hbar}\right)\left[H_{S,}\rho_{S(t)}\right] + \int \mathscr{K}\,(t-\tau)\,\rho_{S(\tau)d\tau}$$

The memory kernel $\mathcal{K}(t - \tau)$ encodes how past system–environment correlations act back on the system. The Markovian limit is recovered when $\mathcal{K}(t - \tau) \propto \delta(t - \tau)$, collapsing the integral to a time-local generator.

Hierarchical Equations of Motion (HEOM)

HEOM replaces the single nonlocal equation by a hierarchy of coupled, time-local equations for the system density operator ρ_0 and a ladder of auxiliary density operators ρ_1, ρ_2, ... that encode successively higher-order environmental memory. A schematic form is:

$$\frac{d\rho}{dt} = -\left(\frac{i}{\hbar}\right)[H_S, \rho] - i \sum V \rho_{\{1,j\}}$$

$$\frac{d\rho_{\{1,j\}}}{dt} = -\left(\gamma + i\mathcal{L}_S\right)\rho_{\{1,j\}} - i\,c\,\left(V\rho - \rho V\right) - i \sum V \rho_{\{2,jk\}}$$

... continues up the hierarchy, truncating when memory is negligible.

Each auxiliary operator represents one layer of environmental memory. Couplings between layers capture feedback ('backflow') of coherence from the environment to the system. Numerically, HEOM is highly accurate for structured baths (e.g., Drude–Lorentz, underdamped modes) and has become a standard in quantum chemistry, quantum transport, and quantum biology.

Morphology: Nesting and Networking Made Explicit

- Nesting (temporal): In the nonlocal form, the integral over past $\rho_{S(\tau)}$ makes the time-nesting explicit. In HEOM, nesting becomes a stack of ADOs (ρ_1, ρ_2,...), each a completion-within-completion.

- Networking (channels): Multiple system operators V_j (or Lindblad-like channels) couple different bath modes and memory paths, forming a distributed network of coherence exchange.
- Markovian limit: When memory decays rapidly (large γ_j), the hierarchy collapses and the dynamics reduce to an effective Lindblad form.

VGF Correspondence

Concept	Time-Nonlocal / HEOM Structure	VGF Interpretation
Temporal nesting	Kernel integral $\int \mathcal{K}(t-\tau)\rho(\tau)\,d\tau$; ADO hierarchy ($\rho_1$, ρ_2,...)	Completions within completions in time
Networking	Multiple V_j channels and inter-ADO couplings	Distributed redundancy and feedback links
Backflow	Non-monotone divisibility; coherence returns	Fidelity re-injection from environment
Markov limit	Short memory → hierarchy truncates	Collapse to local completion (classical appearance)
Steady state	Fixed point of hierarchy	Pointer-like attractor under sustained coupling

Memory, Nesting and Networking

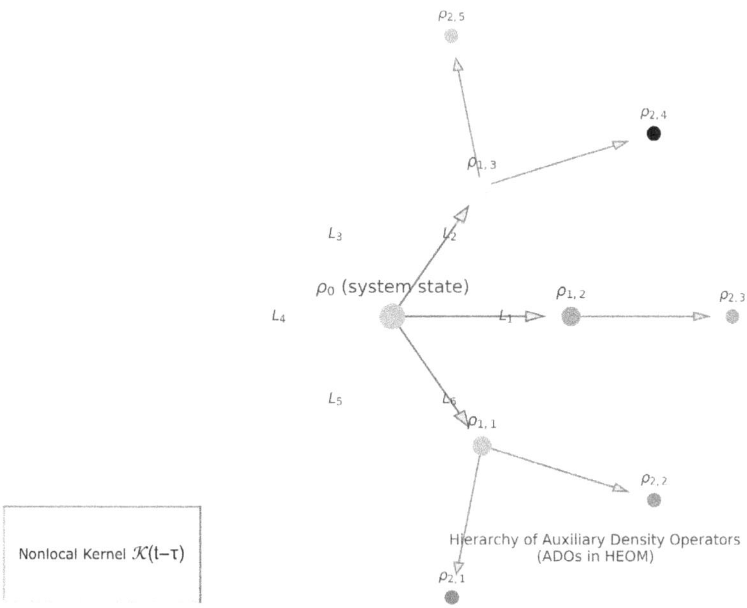

Figure: The system density ρ_0 couples to multiple environmental channels (labels L_k) and to successive memory layers (ρ_1, ρ_2,...). The nonlocal kernel $\mathcal{K}(t-\tau)$ expresses temporal nesting; the interconnections among layers represent networking of memory paths. In the Markov limit, the hierarchy collapses to a single time-local generator (Lindblad).

Review

We see here an important feature of the HEOM - how **it embodies the emergence of *networks* within the *nesting principle*.** The nonlocal kernel $\mathcal{K}(t-\tau)$ carries the nonlocal temporal nesting, with networks of memory paths between nested layers. This morphology is also found in the IIP-VGF approach.

The Environment As Memory

In open quantum systems, the environment retains correlations with the system over finite timescales. This persistence of correlation is called environmental memory or non-Markovianity. Rather than assuming an environment that immediately forgets (as in Markovian models such as the Lindblad equation), the HEOM formalism explicitly represents these memory effects through a hierarchy of coupled density operators. Each level in the hierarchy carries residual coherence from earlier system–environment interactions.

From Memory Kernel to Hierarchical Structure

The starting point is the time-nonlocal master equation:

$$\frac{d\rho_S(t)}{dt} = -\frac{i}{\hbar}[H_S, \rho_S(t)] + \int_0^t \mathcal{K}(t-\tau)\,\rho_S(\tau)\,d\tau\,.$$

Here, the kernel $\mathcal{K}(t-\tau)$ encodes how past interactions influence the present. The HEOM formalism rewrites this equation in a time-local form by expressing the environmental correlation function $C(t)$ as a sum of exponentials:

$$C(t) = \Sigma_j\, c_j\, e^{-\gamma_j t},$$

where each term corresponds to an environmental mode j with memory amplitude c_j and decay rate γ_j. Each exponential generates its own auxiliary density operator (ADO), producing a hierarchy that encodes memory at successive depths.

The Hierarchy as a Memory Structure

At the top level (ϱ_0), the system's reduced density operator evolves as:

$$\frac{d\rho}{dt} = -\left(\frac{i}{\hbar}\right)\left[H_{S,}\rho_0\right] - i\ \Sigma_j\ V_j\ \rho_{\{1,j\}}.$$

Each $\rho_{\{1,j\}}$ represents the system's correlation with environmental mode j - its first-order memory. Subsequent levels $(\rho_{\{2,jk\}}, \rho_{\{3,jkl\}}, \ldots)$ represent correlated or higher-order memories, with shorter lifetimes governed by the damping constants γ_j. In this way, the environment is not a single reservoir but a hierarchy of dynamically coupled memory nodes.

Physical and Mathematical Interpretation

HEOM Level	Physical Meaning	Role as Memory	Lifetime (γ)
ϱ_0	System's present reduced state	Receives feedback from memory layers	—
$\varrho_{1,j}$	Single-mode environmental correlation	Stores first-order memory	$1/\gamma_1$
$\varrho_{2,jk}$	Two-mode correlated memory	Retains higher-order correlations	$1/\gamma_2 < 1/\gamma_1$
$\varrho_{3,jkl}$	Higher-order environmental memory	Stores faint echoes of past coherence	$1/\gamma_3 \ll 1/\gamma_1$
…	Deeper layers	Increasingly short-lived traces	Rapid decay

Flow of Coherence and Backflow

Information and coherence flow down the hierarchy from ϱ_0 into the deeper layers as the system interacts with its environment. When environmental correlations persist, the deeper layers can in turn influence the upper ones, creating backflow—temporary restoration of coherence or population. The strength and frequency of this backflow are controlled by the decay rates γ_j and coupling constants c_j. In the Markovian limit ($\gamma_j \to \infty$), all memory layers decay instantly, and the hierarchy collapses to a single Lindblad equation.

VGF and Stability–Fidelity Interpretation

From the perspective of the Vast Generative Field (VGF), each level in the HEOM represents a local completion—an emergent stabilisation of coherence within the generative continuum. The deeper layers correspond to earlier completions that retain partial fidelity to the generative process. The coupling between layers expresses the dynamic trade-off between stability (environmental redundancy) and fidelity (residual coherence). As the hierarchy deepens and γ increases, stability dominates, and fidelity is lost—culminating in the fully classical regime.

Hierarchy of Environmental Memory in HEOM

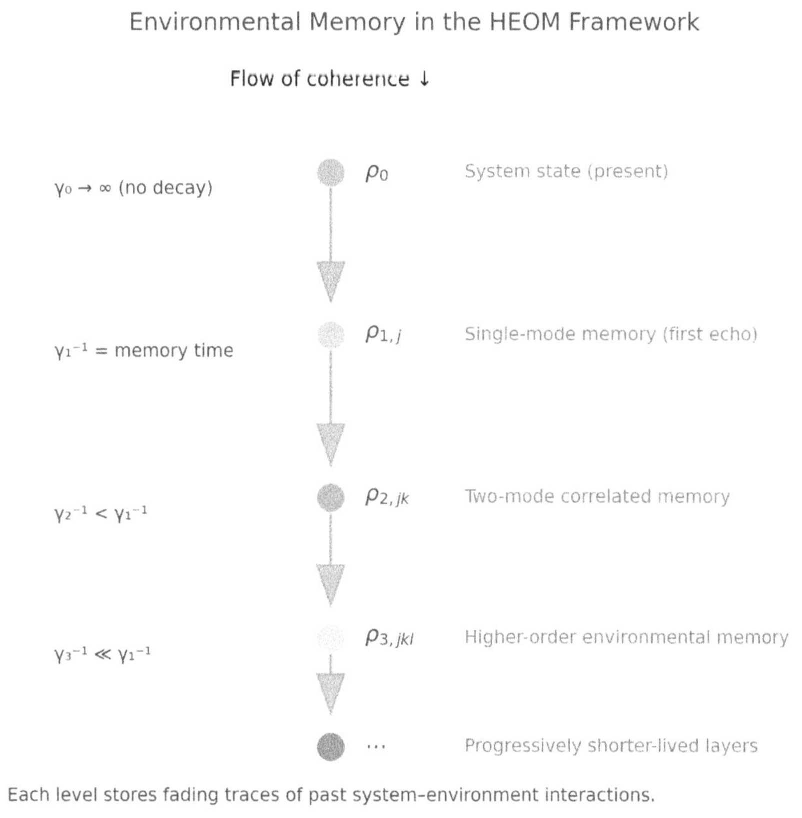

Figure: Each level of the HEOM hierarchy (ρ_0, ρ_1, ρ_2, ρ_3, ...) represents a progressively deeper layer of environmental memory. Arrows denote the flow of coherence and feedback through the hierarchy. Lifetimes decrease with increasing depth (larger γ_j), and the hierarchy truncates when memory becomes negligible, recovering the Markovian limit.

Environment As Memory

The Hierarchical Equations of Motion (HEOM) formalism represents a conceptual shift in how the environment is treated in open quantum systems. In standard Markovian or perturbative models, the environment is regarded as a background reservoir that merely includes memory effects through correlation times. In the HEOM framework, however, the environment is not a separate entity that possesses memory — it is the memory itself, explicitly written in operator form.

Environment Including Memory (Conventional View)

In traditional approaches to open-system dynamics, the environment is viewed as a large collection of degrees of freedom — oscillators, phonons, solvent modes — that retain correlations for finite times. Memory appears as a feature of the environment's behaviour: the correlation function $C(t)$ describes how long an environmental perturbation continues to influence the system. In this sense, the environment includes memory, but remains an external reservoir.

Environment as Memory (HEOM View)

In the HEOM formulation, the environment is expressed directly as a hierarchy of memory operators. Each auxiliary density operator (ADO) represents a distinct memory channel derived from the decomposition of the environmental correlation function:

$$C(t) = \Sigma_j \, c_j \, e^{-\gamma_j t}.$$

Each exponential term defines a memory timescale γ_j^{-1} and a memory amplitude c_j. The associated ADO $\rho_{1,j}$ is not an external object but a record of the system's interaction history with mode j. Higher-order operators ($\rho_j^{(1)}, \rho_{jk}^{(2)}, \rho_{jk\ell}^{(3)}, \rho_{jk\ell m}^{(4)}, \ldots$) represent deeper layers of this memory, corresponding to correlated or repeated interactions. Thus, the entire environment is reconstructed as a structured field of memory—an evolving network of retained information.

The Significance

Treating the environment as memory rather than as an external bath redefines open-system dynamics. The ADOs are not passive recorders — they are active participants in the evolution of coherence and decoherence. Information and coherence flow down the hierarchy into deeper memory layers, while partial backflow restores fidelity at higher levels. The total dynamics become the motion of information through a nested memory architecture, not the exchange of energy with a physical background.

Morphological and VGF Interpretation

In the Vast Generative Field (VGF) perspective, the HEOM hierarchy reveals the morphology of generative recursion: each level is a local completion retaining some fidelity to the generative coherence above it. The environment as memory is therefore the field of iterative stabilisations—the distributed record of generative transformations.

Stability increases and fidelity decreases with memory depth, corresponding to the loss of generative coherence over time. The environment, in this sense, is the **temporally extended memory of the system's own generativity**.

As such, its morphology harmonises completely with the IIP-VGF approach and framework.

From Quantum Memory to Classical Law

The Hierarchical Equations of Motion (HEOM) represent the environment as a structured hierarchy of memory layers (auxiliary density operators, ADOs). Coherence flows from the system's present state (ρ_0) into deeper memory layers (ρ_1, ρ_2, ...), where phases gradually cancel and amplitudes stabilise. The accumulated imprint of these layers forms a web of stable correlations among observables — what we recognise as the deterministic laws of classical physics. This transition is effectively irreversible (stability dominates, fidelity is lost), although reversibility remains implicit at the global, fully coherent level.

Chain of Memory → Web of Classical Relations

Each ADO retains a partial record of prior system–environment interactions. As time passes, the chain of ADOs encodes overlapping temporal correlations across multiple timescales. When coherence decays below experimental detectability, the surviving pattern is a stable network of correlations. This network defines the effective constraints among classical observables—the operational content of classical laws.

Formal Transition

In the deep hierarchy limit, the effective dynamics of ρ_0 reduce to a Markovian master equation (Lindblad type):

$$\frac{d\rho_0}{dt} = -\left(\frac{i}{\hbar}\right)\left[H_{S,}\rho_0\right] + \mathscr{D}\left[\rho_0\right].$$

Here $\mathscr{D}\left[\rho_0\right]$ encodes the cumulative effect of the hierarchy as fixed dissipative and decohering terms. The microscopic history (nested feedback via ADOs) is compressed into macroscopic invariants (classical regularities).

Irreversibility and the Stability–Fidelity Law

Globally the dynamics remain unitary, but as information diffuses into a large environment, it becomes practically unrecoverable. Thus classicality appears as an attractor of maximal stability (redundancy) and minimal accessible fidelity (lost phase information). In the Stability–Fidelity reading: stability ↑ as redundancy ↑; fidelity ↓ as generative coherence is dispersed.

VGF Mapping

Within the Vast Generative Field (VGF), each ADO level is a local completion — a stabilised trace of the generative process. The full chain forms a recursive scaffold whose stable limit is the classical web of relations. Classical laws are thus the most stable configurations producible by the generative field: the fixed-point morphology of nested coherence under sustained coupling.

Summary: Generative Regimes and Outcomes

Regime	Physical Analogue	Dominant Feature	Outcome
Infinite coherence (VGF)	Pure quantum potential	No partitions, maximal fidelity	Unstable unity
Nested ADO hierarchy	Open-system dynamics	Iterative stabilisations	Emergent decoherence
Deep hierarchy / redundancy saturation	Thermodynamic limit	Maximal stability, near-irreversibility	Classical lawfulness
Truncated hierarchy	Finite-memory devices	Residual coherence, partial reversibility	Quantum–classical boundary

ADO Chain → Classical Correlation Web

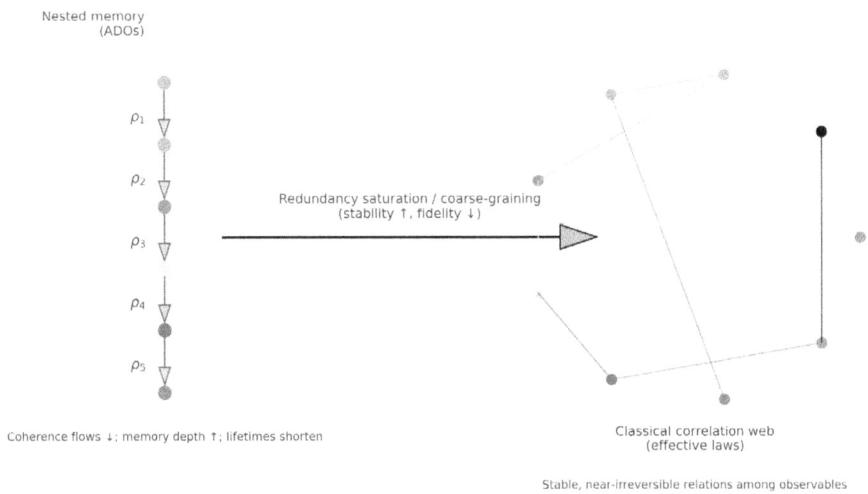

Figure: **Left**: the vertical chain of ADOs (ρ_0, ρ_1, ρ_2, ...) encodes nested memory and feedback. **Right**: as phases cancel and redundancies accumulate, the hierarchy compresses into a flattened network of stable classical correlations—the empirical laws linking observables. The horizontal arrow represents coarse-graining and redundancy saturation (stability ↑, fidelity ↓).

Recursive Memory in the HEOM

The Hierarchical Equations of Motion (HEOM) are iterative in essence, but not written in a way that looks like a simple recursive mapping such as

$$x_{n+1} = f(x_n)\,.$$

They're expressed as coupled differential equations, not as discrete-step updates — yet the structure of the hierarchy is already recursive.

The HEOM are Essentially Continuous Recursion

Each tier of the hierarchy,

$$\dot{\rho}^{(n)} = \mathscr{L}_S\rho^{(n)} - \sum_j \gamma_j n_j \rho^{(n)} \bullet i \sum_j [V, \rho_j^{(n+1)}] \bullet i \sum_j n_j (c_j V \rho_j^{(n-1)} - c_j^* \rho_j^{(n-1)} V)$$

depends on:

- the next higher tier ($\rho^{(n+1)}$), and
- the next lower tier ($\rho^{(n-1)}$).

This is literally recursion in continuous time:
$\rho^{(n)} \longleftrightarrow F(\rho^{(n+1)}, \rho^{(n)}, \rho^{(n-1)})$.
So the equation is recursive — just not in discrete-step form.

Each tier is an iteration in the correlation order between system and environment. It is effectively **time-continuous recursion across correlation depth**.

Making Iteration Explicit

If we want to see the iterative nature explicitly, we can rewrite the HEOM schematically as a mapping operator $\mathscr{F}$ acting on the hierarchy vector:

$$\dot{\mathbf{R}} = \mathscr{F}(\mathbf{R}),$$

where

$$\mathbf{R}(t) = \begin{bmatrix} \rho^{(0)}(t) \\ \rho^{(1)}(t) \\ \rho^{(2)}(t) \\ \vdots \end{bmatrix}.$$

The hierarchy operator $\mathscr{F}$ is block-recursive — each block couples a level to its neighbours.

Then, integrating in small time steps Δt:

$$\mathbf{R}_{k+1} = \mathbf{R}_k + \Delta t \, \mathscr{F}(\mathbf{R}_k),$$

we see the iteration explicitly.

This is exactly how HEOM is implemented computationally: each step updates the hierarchy recursively, tier by tier, until convergence.

So numerically, HEOM is already realised as an iterative recursion — it is just that it isn't usually presented this way in analytic form.

"Temporal Recursion" *vs* "Correlation Recursion"

There are two iterative dimensions built into the HEOM:

Iteration type	Variable	Physical meaning
Temporal iteration	$t \to t + \Delta t$	Continuous time evolution (integration of differential equation)
Correlational iteration	$n \to n + 1$	Recursion in systemâ€"environment correlation order

Thus the HEOM is a 2D recursion scheme:

$$\rho^{(n)}(t + \Delta t) = \rho^{(n)}(t) \bullet \Delta t\, f\big(\rho^{(n-1)}(t), \rho^{(n)}(t), \rho^{(n+1)}(t)\big).$$

So the iteration happens:

- horizontally across time, and
- vertically across correlation tiers.

This is a **structural mirror of VGF morphology**: timewise iteration (looping) nested within hierarchical recursion (nesting of loops).

Explicit Recursion Form

We can define an iteration functional $\mathcal{I}$ such that:

$$\rho^{(n)}(k + 1) = \mathcal{I}\big[\rho^{(n)}(k);\ \rho^{(n-1)}(k), \rho^{(n+1)}(k)\big].$$

Then the evolution becomes:

$$\rho^{(n)}(k) \longrightarrow \rho^{(n)}(k+1) = \rho^{(n)}(k) - i\Delta t\big(\mathscr{L}S\rho^{(n)}(k) + \sum_j [V, \rho^{(n+1)}(k)] + \dots\big)$$

This shows the recursive dependence explicitly — akin to an infinite iteration process (IIP) expressed over nested levels.

Now the convergence of the HEOM hierarchy — where $\rho^{(n)} \to 0$ for high n and $\rho^{(0)}$ stabilises — can be read as an iterative fixed point:

$$\rho^{(0)} = \mathscr{I}\big[\rho^{(0)};\, \rho^{(1)}, \rho^{(-1)}\big] \quad \text{with} \quad \rho^{(n>0)} \to 0.$$

So the steady-state density operator is literally the fixed point of a recursive iteration between system and environment tiers.

This, precisely, is a mathematical image of **loop closure** in the VGF picture.

Morphological Interpretation

HEOM form	IIP–VGF equivalent
Coupled hierarchy of $\rho^{(n)}$	Nested field of local iterations
Recursion across correlation order	Relations between relations of iterations
Convergence to steady state	Local completion (loop closure)
Differential operator $\mathscr{L}_S$	Internal iteration generator
Memory kernels (γ, c_j)	Fidelity decay terms (redundancy buildup)

So even though HEOM is written as continuous dynamics, it is fundamentally recursive — an infinite iterative structure in correlation depth and in time, except that for specific purposes it may be truncated.

Review

Recursive Morphology of the Hierarchical Equations of Motion (HEOM)

Although the HEOM is expressed as a continuous-time differential hierarchy, its structure is intrinsically recursive. Each tier's evolution depends on the tiers immediately above and below, forming a continuous iteration across correlation depth.

When integrated over time, the equations become explicit recursions in both time and hierarchy order, converging toward a fixed-point attractor — the decohered steady state.

Thus, the HEOM is already an iterative system in two dimensions: a continuous realisation of infinite iteration processes within a nested field of relations, exactly mirroring the morphology of the VGF.

Iteration Dynamics

The Hierarchical Equations of Motion express open quantum dynamics as a recursive flow between tiers of system–environment correlation. Although formally differential, the HEOM form a two-dimensional iteration structure:

- Horizontally, each density operator evolves in time through infinitesimal steps of unitary and dissipative motion.
- Vertically, each tier $\rho^{(n)}$ couples to those immediately above and below — $\rho^{(n+1)}$ and $\rho^{(n-1)}$ — representing progressively deeper orders of memory and correlation between the system and its environment.

Together these axes create a continuous recursion lattice, where time evolution and correlation recursion interlock. At every moment, information flows both forward in time and upward through correlation depth, while coherence decays downward toward the base tier $\rho^{(0)}$.

Convergence and Stabilisation

As higher-order tiers relax ($\rho^{(n>0)} \rightarrow 0$), their feedback to the base tier diminishes. The lowest level, $\rho^{(0)}$, then converges toward a steady-state attractor — the decohered or thermally equilibrated system density operator. Mathematically, this is the fixed point of the recursion; conceptually, it is the closure of generative loops: coherence condenses into redundancy, forming a stable pointer-state domain.

In this view, the HEOM is not merely a numerical tool but a morphological map of nested iteration. Each **vertical** recursion step corresponds to a higher-order stabilisation of coherence; each **horizontal** step corresponds to temporal iteration. The final convergence of $\rho^{(0)}$ thus represents the local completion of an infinite iterative process — an attractor emerging from continuous generative recursion.

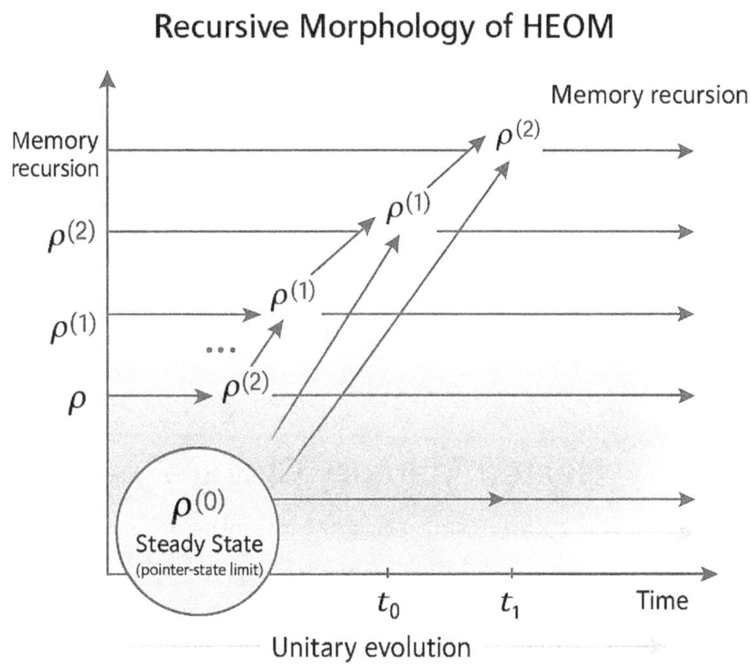

Two-dimensional iteration in the HEOM: horizontal (temporal) and vertical (correlational). Convergence of all paths leads to stabiliszed system density operator.

Figure: Coherence decays downwards towards the base tier. Diagonal arrows point upwards towards where memory translated into the steady state, has come from. As higher order tiers relax in time their feedback to the base tier diminishes. The lowest level then converges towards the decohered system density operator (bottom circle). Each tier "remembers" a layer of past correlations from above. When all these tiers reach dynamical balance — the top terms decaying as fast as they feed back down — the recursion closes on itself at infinite recurrence.

The steady state still contains the entire nested causal history, now compressed into effective rates and correlations. So the bottom circle (pointer ensemble) is a memory-saturated fixed point: all higher tiers have fed their information into it. Memory is not lost, rather, it is absorbed into structure. Therefore the classical "pointer state" is the stable residue of infinite nested iteration — the memory of coherence translates to persistent relational form.

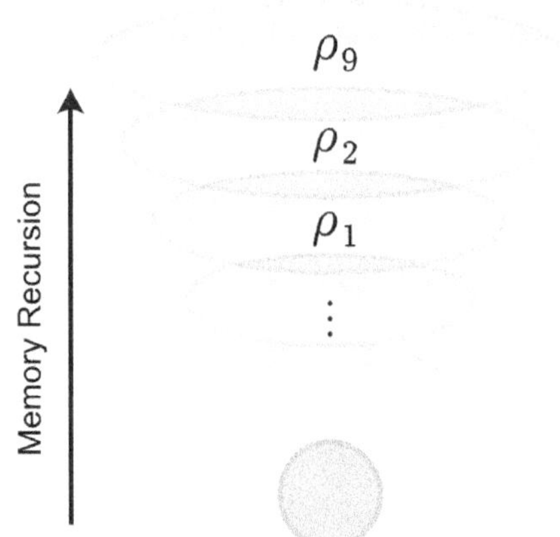

Nested Memory Closure

The steady-state pointer ensemble is formed from the nested memory levels in the HEOM.

The Partition in the HEOM

In Chapter 2 we discussed the significance of the **partition** between "system" and "environment", in the understanding of decoherence. We said that in the case of standard quantum decoherence theory, and quantum Darwinism, the density matrix has to be partitioned and the environment "traced out" in order to arrive at the representation of the decohered state of the system.

Even in the case of the Hierarchical Equations of Motion it is necessary to manually introduced a partition in the model. However, in the case of the HEOM, the dynamics then go on to embody the description of a partition that is *emergent*.

The Technical Starting Point

In physics, the HEOM is derived from a system–bath Hamiltonian, written explicitly as

$$H_{\text{tot}} = H_S + H_B + H_{SB}.$$

Here:

H_S is the "system" Hamiltonian, describing the degrees of freedom of interest (e.g. an exciton, a spin, a molecular mode);

H_B is the "bath" or environment, modelled as a large collection of harmonic oscillators or modes;

H_{SB} is their interaction, often in bilinear form $H_{SB} = V \sum_j c_j x_j$, coupling a system operator V to bath coordinates x_j.

At this formal level, the partition has already been imposed "by hand". It's how the physicist defines what counts as "the system" — what we track — versus "the environment" — what we trace out.

This is the first and most literal meaning of the partition in HEOM.

What the HEOM Then Does

Once the split is made, the total density matrix ρ_{tot} evolves unitarily under the Liouville–von Neumann equation

$$\dot{\rho}\,\text{tot} = -\frac{i}{\hbar}[H\text{tot}, \rho_{\text{tot}}].$$

But because we can't keep track of the bath's exponentially large Hilbert space, we "trace it out":

$$\rho_S(t) = \text{Tr}B[\rho\,\text{tot}(t)].$$

This tracing procedure — the partial trace — mathematically **enacts** the system–environment split.

However, it doesn't eliminate the environment: its influence reappears as memory kernels in the evolution of ρ_S.

The HEOM expresses those kernels as the hierarchy of Auxiliary Density Operators (ADOs), each representing a different level of correlation between system and environment.

So while the initial partition is imposed, the hierarchy itself reconstructs the boundary dynamically through the flow of correlations between the two sides.

How the Partition Becomes Emergent

In the HEOM, the coupling between tiers has the general form

$$\dot{\rho}^{(n)} = \mathscr{L}_S \rho^{(n)} + \sum_j \mathscr{A}_j \rho_j^{(n+1)} + \sum_k \mathscr{B}_k \rho_k^{(n-1)}.$$

Each level n corresponds to system–environment correlations of order n.

At n=0, we have the "bare system".

At higher n, we include successively more of the environment's structure.

When correlations at a given order decay faster than the timescale of interest, they can be truncated — that truncation defines the effective partition.

Thus, the location of the system–environment split becomes scale-dependent:

- At high temporal resolution, the "system" must include more ADO tiers (more of its environment).
- At coarse scales, many ADOs can be neglected, and the system looks effectively isolated.

In this sense, the partition emerges from decoherence itself — it is not fixed, but rather, it is dynamically reconstructed at each level of stabilisation.

The VGF Interpretation

In VGF terms, this corresponds to **local completion**:

In HEOM	In the VGF
Each truncation of the hierarchy defines a temporarily closed	Each local stabilisation defines a temporarily closed loop
The rest of the hierarchy acts as its "environment"	The surrounding field acts as its generative context
Memory exchange across tiers maintains coupling	Iterative feedback maintains coherence across scales
Boundary (partition) moves depending on timescale and coherence length	Boundaries (partitions) emerge wherever coherence locally condenses into redundancy

So the HEOM embodies the idea that the system/environment partition is not **fundamental** — rather, it is **a dynamic consequence of stabilisation**, precisely as in the VGF.

Review

In the HEOM formalism, the system–environment partition begins as an explicit assumption but becomes a dynamic, scale-dependent construct.

The lowest tier describes the system; higher tiers encode the environment's memory and correlations.

As correlations decay or are truncated, the effective boundary shifts, producing a nested hierarchy of temporary "systems" each with its own environment.

In the VGF framework, this is the mathematical realisation of local completion: the emergence of relatively closed structures (loops) within an open, generative field.

The partition is thus not fundamental—it is the moving interface where coherence condenses into structure.

The Emergent Partition: System–Environment Boundaries in the HEOM and VGF Frameworks

In traditional quantum mechanics, the system–environment split is defined a priori. The Hilbert space is factorised into system and environment components ($H = H_S \otimes H_E$), and the system's state is obtained by tracing out the environment. This fixed partition is adequate for Markovian approximations, but it fails when memory and feedback are significant. In the Hierarchical Equations of Motion (HEOM) framework, and in the broader Stability–Fidelity perspective, the partition is no longer static: it emerges dynamically as coherence flows and decoheres through successive levels of interaction.

Fixed vs. Emergent Partitions

Fixed Partition (Markovian models):

The boundary between system and environment is predefined and remains constant. All environmental influence is encoded through phenomenological decay rates.

Emergent Partition (HEOM / Non-Markovian):

The distinction between system and environment evolves dynamically. Each auxiliary density operator (ADO) represents a layer of interaction in which parts of the environment retain coherence and act as extensions of the system itself.

The Partition as a Decoherence Gradient

In HEOM, each hierarchy level (ρ_0, ρ_1, ρ_2, ...) corresponds to a different degree of coherence retention. Rather than a single trace-out operation, decoherence occurs gradually: coherence leaks into environmental modes, which remain correlated with the system for finite times before fading. Each level therefore defines a temporary, scale-dependent boundary—a local completion separating a coherent core from its redundant surround.

Physical and Mathematical Interpretation

At level ρ_0, the 'system' consists of degrees of freedom still maintaining coherence. At level ρ_1, the system's influence has extended into the environment's first-order correlations; the boundary shifts outward. At deeper levels (ρ_2, ρ_3, ...), coherence is nearly lost, but traces of correlation persist. The partition thus emerges from the degree of correlation retained, rather than from any ontological distinction between system and environment.

VGF Mapping and Morphology

In the Vast Generative Field (VGF) framework, each emergent partition corresponds to a surface of stability formed within the generative flux. The 'system' is a region of coherence stabilised through redundancy; the 'environment' is the surrounding field of stabilisations that feed back and constrain it. Each new partition represents a loss of fidelity and an increase in stability—an expression of the Stability–Fidelity Law. The boundaries $\rho_0, \rho_1, \rho_2, \ldots$ are thus nested stability surfaces arising from iteration within the generative continuum.

Summary

Regime	Description	Nature of Partition	VGF Interpretation
Global Coherence	System and environment indistinguishable	No partition	Pure generative unity
Weak Coupling	Partial entanglement, temporary boundaries	Oscillatory partition	Emergent local completion
Non-Markovian (HEOM)	Hierarchy of correlated layers	Dynamic, scale-dependent	Nested completions with feedback
Markovian (Lindblad)	Environment as memoryless reservoir	Fixed, imposed boundary	Static stabilisation
Classical Limit	Complete decoherence, redundant copies	Persistent but approximate partition	Fully stabilised objectivity

Emergent System–Environment Boundaries

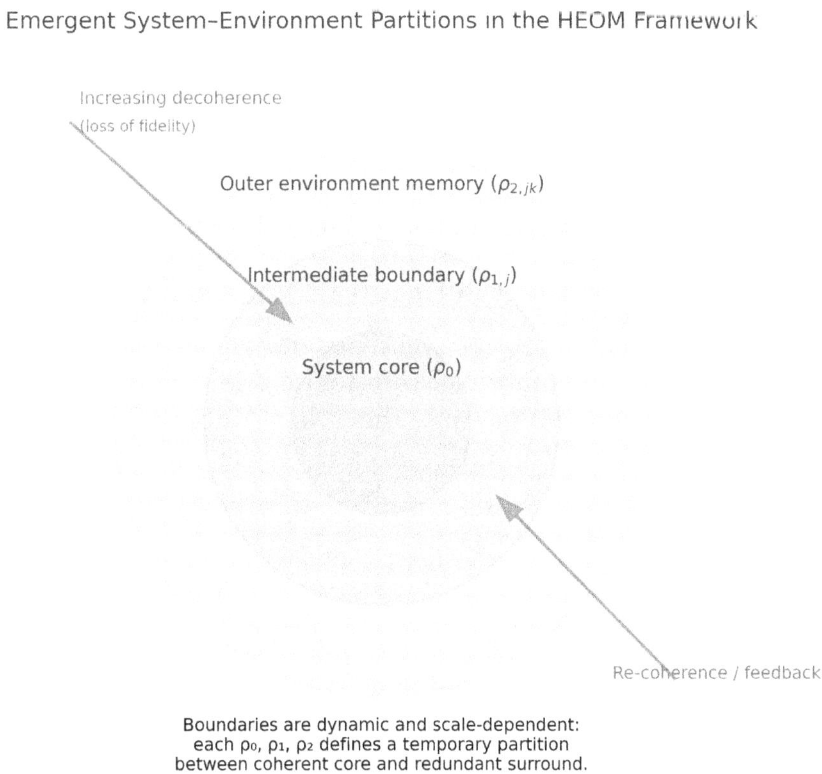

Figure: Nested system–environment boundaries emerging dynamically in the HEOM framework. The inner region (ρ_0) represents the coherent system core; surrounding layers (ρ_1, ρ_2) represent successive levels of partial decoherence and environmental memory. Arrows indicate the gradient of decoherence (loss of fidelity) and possible re-coherence feedback. Each boundary is temporary and scale-dependent, corresponding to a local completion in the VGF morphology.

Fourier Duals

Fourier Duals as Footprints of Iteration

Attractors from Infinite Iteration Processes (IIPs)

At the most abstract level, the Vast Generative Field (VGF) can be thought of as an infinite fabric of Infinite Iteration Processes (IIPs) — recursive generative dynamics that, left unbounded, produce structures at all scales. These IIPs do not yet contain "objects" in the ordinary sense; they are pure coherence, an undivided sea of potential.

Yet within such a field, stable attractors naturally arise: loops of iteration that "catch" and repeat indefinitely. Numbers themselves can be interpreted this way: archetypal stabilisations of iteration. The number 2, for example, is not a mere symbol, but a stable attractor form in the VGF that manifests redundantly in many guises (written symbol, quantity of apples, eigenvalue in an equation).

From the perspective of the Stability–Fidelity Law, these attractors embody a trade-off:

- Stability is gained by locking iteration into reproducible loops.
- Fidelity to the deeper generative coherence is lost — each stabilised attractor is only a thin residue of the infinite process that gave rise to it.

Decoherence Overlayed onto IIP Attractors

Quantum mechanics provides a ready-made model of this stabilisation. A global pure state $|\Psi\rangle$ encodes coherence across system and environment. When we trace out the environment, interference terms vanish and the system is described by a mixed density matrix:

$$\rho = \Sigma_i \, p_i \, \left|u_i\right\rangle \left\langle u_i\right|.$$

Here the outer products $\left|u_i\right\rangle \left\langle u_i\right|$ are projectors onto pointer states. They are the mathematical crystallisation of attractors: stable forms that survive entanglement and redundancy.

From the IIP–VGF viewpoint, decoherence is exactly the translation of fidelity into stability:

- The original IIP-driven coherence (superposition) retains maximum fidelity.
- Redundant coupling to the environment drives states toward orthogonality, making them distinguishable and repeatable.
- These orthogonal pointer states are the stabilised attractors, directly parallel to the way numbers or equations emerge as redundant residues of generative iteration.

Thus: decoherence in physics is the operational manifestation of the same law that governs how attractors stabilise in the VGF.

Fourier Duals as Footprints of IIPs

(a) Position and Momentum

The inner product between position and momentum eigenstates is:

$$\langle x | \, p \rangle = \left(\frac{1}{\sqrt{2\pi\hbar}}\right) e^{i\,p\,x/\hbar}.$$

This exponential kernel is nothing other than an infinite iteration in the complex plane: $e^{i\theta} = cos\theta + isin\theta$, endlessly rotating.

The Fourier transform expresses the wavefunction in either basis:

$$\psi\,(x) = \left(\frac{1}{\sqrt{2\pi\hbar}}\right) \int e^{i\,p\,x/\hbar} \phi\,(p)\,dp,$$

$$\phi(p) = \left(\frac{1}{\sqrt{2\pi\hbar}}\right)\int e^{-i\,p\,x/\hbar}\psi(x)\,dx.$$

This duality shows the IIP footprint: oscillatory iteration is the bridge between the two representations.

- Position localisation (sharp ψ(x)) requires superposition of many oscillations — fidelity to spatial detail, but instability in momentum.
- Momentum localisation (sharp ϕ(p)) is a pure oscillation spread across all space — stability in momentum, but loss of fidelity to position.

This is Heisenberg's uncertainty principle:

$$\Delta x\ \Delta p\ \geq \frac{\hbar}{2},$$

now seen as the Stability–Fidelity Law written in Fourier space.

(b) Energy and Time

Similarly, the kernel between energy and time representations is:

$$\langle t|\,E\rangle\ \propto\ e^{-\frac{iEt}{\hbar}}.$$

Energy eigenstates evolve as pure phase oscillations:

$$\psi(t) \sim e^{-iEt/\hbar}.$$

Again, this exponential is an IIP trace — infinite iteration in phase.

- Sharp energy eigenstate: perfectly stable number attractor, but delocalised in time (fidelity lost).
- Localised temporal event: requires superposition of many energies, preserving fidelity to momentary detail, but unstable in energy.

The energy–time uncertainty principle

$$\Delta E \, \Delta t \, \gtrsim \frac{\hbar}{2},$$

is the exact analogue of the position–momentum case.

Unified Picture

- **VGF coherence**: infinite IIPs, maximum fidelity, no stability.
- **Attractors (numbers, pointer states)**: stabilised residues of iteration, redundancy-rich, fidelity-poor.
- **Decoherence**: the process that enforces stability by driving overlaps into orthogonality.
- **Fourier duals**: the mathematical footprint of IIPs, showing explicitly how stabilisation in one domain requires fidelity-loss in its conjugate domain.

Condensed Principle

The wave equation, the Fourier kernels, and the uncertainty relations are direct signatures of infinite iteration processes in physics. Decoherence is the mechanism that overlays stability onto fidelity, turning the raw coherence of IIPs into stabilised attractors. Position–momentum and energy–time dualities bear the unmistakable footprint of this iterative generativity: oscillatory kernels that mediate the trade-off between fidelity-rich distributions and stability-rich attractors.

The Density Matrix as a Strange Attractor (SNA)

From Iteration to Matrix

Given a quadratic tensor recursion $g(\mathscr{K})$, any stabilisation $\mathscr{K}_*$ induces a linear operator $Dg_{\mathscr{K}*}$ whose matrix representation M encodes the network of couplings between modes of the stabilisation.

So the adjacency matrix of a network can be seen as the linearised shadow of an underlying iterative generative process.

In the quantum setting, the density matrix ρ of a system is an element of the operator space $\mathscr{B}(\mathscr{H})$, placing it within the same mathematical domain as the operator $\mathscr{K}$ in the quadratic tensor recursion

$$g(\mathscr{K}) = \alpha\mathscr{K}^2 + \beta\mathscr{K} + \gamma I.$$

Consequently, any stabilised quantum state ρ_* — whether arising as a fixed point of a discrete CPTP map, a steady state of a Lindbladian generator, or a decohered pointer-state sector — admits a linearisation of the evolution map at ρ_*.

The Fréchet derivative $Dg_{\rho*}$ defines a linear operator on $\mathscr{B}(\mathscr{H})$, whose matrix representation in any finite-dimensional invariant subspace constitutes the adjacency matrix of a dynamical network coupling the relevant populations and coherences.

Thus the evolution of small perturbations $\Delta\rho$ near ρ_* obeys

$$\Delta\rho_{n+1} \approx Dg_{\rho_*}(\Delta\rho_n),$$

providing a canonical transition from the nonlinear generative dynamics of the recursion to a network description of its stabilised modes.

In this sense, the network structure familiar from Liouville-space representations of quantum channels arises naturally as the linearised shadow of the underlying iterative generative process, precisely mirroring the iteration-to-network relationship in the broader IIP–VGF framework.

Formal Framework

Setup

Let $\mathscr{H}$ be a (real or complex) Hilbert space, and let $\mathscr{B}(\mathscr{H})$ denote the space of bounded linear operators on $\mathscr{H}$.

Fix scalars $\alpha, \beta, \gamma \in \mathbb{R}$ or $\mathbb{C}$, and define the quadratic operator map

$g : \mathscr{B}(\mathscr{H}) \to \mathscr{B}(\mathscr{H}), \qquad g(\mathscr{K}) = \alpha\, \mathscr{K}^2 + \beta\, \mathscr{K} + \gamma\, I_{\mathscr{H}}.$

An iterative dynamics on operators is then given by

$$\mathscr{K}_{n+1} = g(\mathscr{K}_n), \qquad \mathscr{K}_0 \in \mathscr{B}(\mathscr{H}).$$

We regard stabilisations as fixed points or attractors of this recursion; here we focus on fixed points.

Lemma (Linearisation and induced network operator)

Lemma.

Let $\mathscr{K}_* \in \mathscr{B}(\mathscr{H})$ be a fixed point of g, *i.e.*

$$g(\mathscr{K}_*) = \mathscr{K}_*.$$

Then the Fréchet derivative of g at $\mathscr{K}_*$,

$$Dg_{\mathscr{K}_*} : \mathscr{B}(\mathscr{H}) \to \mathscr{B}(\mathscr{H}),$$

is the linear operator

$$Dg_{\mathscr{K}_*}(\Delta\mathscr{K}) = \alpha\big(\mathscr{K}_*\Delta\mathscr{K} + \Delta\mathscr{K}\,\mathscr{K}_*\big) + \beta\,\Delta\mathscr{K},$$

for all $\Delta\mathcal{K} \in \mathcal{B}(\mathcal{H})$. In particular, choosing any finite-dimensional subspace

$$\mathcal{S} \subset \mathcal{B}(\mathcal{H})$$

that is invariant under $Dg_{\mathcal{K}*}$ and any basis of $\mathcal{S}$, the restriction

$$Dg_{\mathcal{K}*}\Big|_{\mathcal{S}} : \mathcal{S} \to \mathcal{S}$$

is represented by a square matrix M, which may be interpreted as a network adjacency matrix on the "modes" spanning $\mathcal{S}$.

Proof.

For any $\mathcal{K}$, $\Delta\mathcal{K} \in \mathcal{B}(\mathcal{H})$,

$$g(\mathcal{K} + \Delta\mathcal{K}) = \alpha(\mathcal{K} + \Delta\mathcal{K})^2 + \beta(\mathcal{K} + \Delta\mathcal{K}) + \gamma I_{\mathcal{H}}.$$

Expanding,

$$g(\mathcal{K} + \Delta\mathcal{K}) = \alpha(\mathcal{K}^2 + \mathcal{K}\Delta\mathcal{K} + \Delta\mathcal{K}\,\mathcal{K} + \Delta\mathcal{K}^2) + \beta\mathcal{K} + \beta\Delta\mathcal{K} + \gamma I_{\mathcal{H}}$$

Thus

$$g(\mathcal{K} + \Delta\mathcal{K}) - g(\mathcal{K}) = \alpha\big(\mathcal{K}\Delta\mathcal{K} + \Delta\mathcal{K}\,\mathcal{K}\big) + \beta\Delta\mathcal{K} + \alpha\,\Delta\mathcal{K}^2.$$

The term linear in $\Delta\mathcal{K}$ is

$$L_{\mathcal{K}}(\Delta\mathcal{K}) := \alpha\big(\mathcal{K}\Delta\mathcal{K} + \Delta\mathcal{K}\,\mathcal{K}\big) + \beta\Delta\mathcal{K},$$

and the remainder term is

$$R_{\mathcal{K}}(\Delta\mathcal{K}) := \alpha\,\Delta\mathcal{K}^2,$$

which is $O(\|\Delta\mathcal{K}\|^2)$in operator norm. Therefore $Dg_{\mathcal{K}} = L_{\mathcal{K}}$ is the Fréchet derivative of g at $\mathcal{K}$.

Evaluating at $\mathscr{K} = \mathscr{K}_*$ gives the stated formula.

If $\mathscr{S}$ is finite-dimensional and invariant under $Dg_{\mathscr{K}_*}$, then with respect to any basis $\{e_1, \ldots, e_m\}$ of $\mathscr{S}$, the linear operator $Dg_{\mathscr{K}_*}\Big|_{\mathscr{S}}$ is represented by an $m \times m$ matrix M.

The entry M_{ij} encodes the contribution of mode e_j to the updated perturbation along mode e_i; thus M can be interpreted as an adjacency or coupling matrix on these modes.

Corollary (Iteration ⇒ network form on stabilisation modes)

Corollary.

Let $\mathscr{K}_*$ be a fixed point of the recursion $\mathscr{K}_{n+1} = g(\mathscr{K}_n)$, and let $\mathscr{S} \subset \mathscr{B}(\mathscr{H})$ be a finite-dimensional invariant subspace of $Dg_{\mathscr{K}_*}$ spanned by a distinguished set of perturbation modes (*e.g.* directions corresponding to particular traits, lineages, or interaction patterns). Then:

1. Small deviations $\Delta\mathscr{K}_n \in \mathscr{S}$ from the stabilisation evolve approximately linearly by

$$\Delta\mathscr{K}_{n+1} \approx Dg_{\mathscr{K}_*}(\Delta\mathscr{K}_n),$$

and hence their coordinate vector $x_n \in \mathbb{R}^m$ (or $\mathbb{C}^m$) satisfies

$$x_{n+1} \approx Mx_n,$$

where M is the matrix representation of $Dg_{\mathscr{K}_*}\Big|_{\mathscr{S}}$.

2. The adjacency matrix M thus obtained is the network form induced by the quadratic tensor recursion g on the chosen stabilisation modes. In particular, the pattern of nonzero entries in M captures which modes of the stabilisation dynamically influence which others, making the network description a linearised shadow of the underlying iterative generative dynamics.

The Density Matrix

If we represent the density matrix

$$\rho = \sum_{i,j} \rho_{ij} \, | \, i \rangle \langle j \, | \, ,$$

then:

The nodes i, j correspond to the basis states $| \, i \rangle$.

The edges connecting them are the off-diagonal elements ρ_{ij} (for $i \neq j$).

Each edge carries a complex amplitude encoding both the correlation strength (magnitude) and relative phase (argument).

In a full quantum superposition, every node is linked to every other — a complete directed graph of complex-weighted edges.

That web of connectivity is what coherence is.

What the Edges "Do" Dynamically

Under unitary iteration ($\dot{\rho} = -\frac{i}{\hbar}[H, \rho]$) the edges don't simply fade or persist; they rotate in phase, producing oscillatory coherence patterns.

When we introduce interaction with an environment, each edge also damps — its amplitude decays as information about that phase relation leaks into external degrees of freedom.

So, the decoherence process is literally a progressive loss of off-diagonal edges in this network picture.

Mapping to a Strange-Attractor Form

In a classical strange attractor (say, the Lorenz set or a dissipative map):

- The nodes (sampled points) trace the attractor manifold;
- The edges represent correlations between successive states of the system;
- The geometry of the attractor encodes the invariant correlations that persist under iteration.

In the density-matrix case, the persistent pattern of correlations (which off-diagonal terms survive longest and in what phase relation) is exactly analogous.

The "shape" of ρ in Hilbert space — given by the magnitude/phase topology of its off-diagonals — is a quantum strange attractor living in the complex plane of correlations.

When decoherence proceeds, the system's orbit in this abstract correlation space collapses toward a lower-dimensional manifold (the diagonal): that's the quantum equivalent of an attractor losing its internal loops as dissipation dominates. The link lives in the edges — but also in their collective pattern

So the edges carry the link between the density-matrix representation and the attractor morphology.

However, the attractor is not any single edge: it's the global invariant geometry of the network — the way those edges evolve and self-organise under iteration.

When the environment interacts, it acts like friction plus coarse-graining in the iterative map: energy and phase correlations flow out, collapsing the web onto a smaller attractor set.

The VGF Interpretation

In the VGF frame this becomes beautifully direct:

- The VGF's relations between relations are the origin of these edges;
- The density matrix is a local stabilisation — a snapshot of a self-referential loop network at a given scale;

- Decoherence corresponds to successive local completions that prune the high-fidelity edges (fine-scale coherence) and reinforce redundant loops (stability).

So the connection between the density matrix and its attractor form is entirely encoded in the edges — but the meaning of that connection lies in how the pattern of edges reorganises as iteration proceeds.

Compact Formulation

Edges (ρ_{ij}) carry fidelity.   Diagonals (ρ_{ii}) carry stability.

Decoherence is the iterative re-weighting of this network, driving the system from a fidelity-dominated attractor (rich coherence web) to a stability-dominated attractor (the diagonal and redundant pointer states).

Review

The density matrix represents a network of correlations in Hilbert space. Its off-diagonal elements (edges) encode coherence and phase relations between basis states, while its diagonal elements (nodes) carry the stable probabilities of pointer states. Decoherence corresponds to the iterative damping of edges—an attractor-like contraction of the correlation web into a lower-dimensional manifold. Within the Vast Generative Field (VGF) framework, this process embodies the transformation from fidelity-rich loops to stability-rich structures.

In the VGF, each coherence edge is a relational loop that has not yet closed as a redundancy. Decoherence corresponds to the pruning of these open loops. The density matrix is the attractor landscape of that pruning: a self-limiting network whose topology encodes the transition from quantum coherence to classical stability.

In More Depth

Linear Quantum Evolution and Emergent Nonlinearity

So the density matrix can be viewed as a dynamic network of correlations whose topology evolves under decoherence. Its evolution describes the contraction of a complex coherence web into a lower-dimensional manifold of stable nodes—a process mathematically analogous to the formation of a strange attractor in dynamical systems.

Within the Vast Generative Field (VGF) framework, the density matrix represents **an intermediate generative layer** between pure quantum coherence and classical stability. Its theoretical "decoherence" by partitioning and "tracing out the environment" only represents the first stage of natural decoherence, up to its representation of the "measurement problem".

The Density Matrix as a Coherence Network

In Hilbert space, each basis state $|i\rangle$ corresponds to a node, and each off-diagonal term $\rho_{i,j}$ represents a complex-weighted edge linking nodes i and j. Hermiticity enforces reciprocity between edges, while unit trace ensures global conservation of probability. The density matrix is thus a coherent adjacency network whose evolution under the Liouvillian operator $\mathcal{L}$ is a continuous reweighting of edges.

As decoherence proceeds, this network becomes sparser: off-diagonal edges fade, leaving diagonals dominant. This contraction toward a lower-dimensional manifold corresponds to the emergence of stable pointer states.

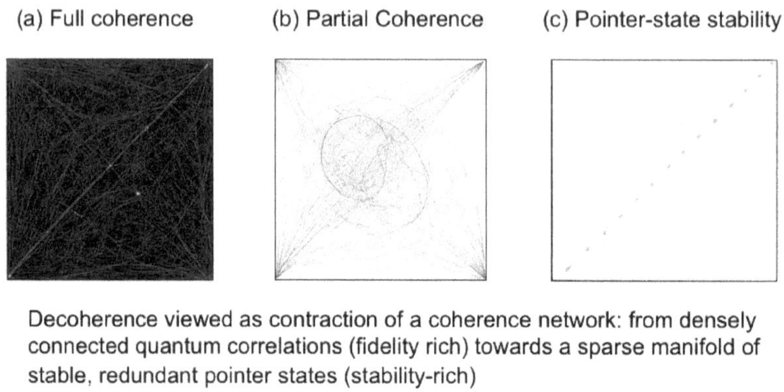

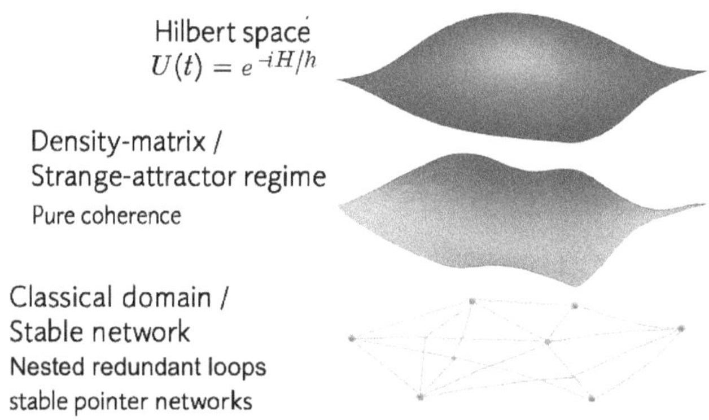

Figure: Decoherence as contraction of a coherence network. The density matrix occupies the intermediate generative layer: a strange attractor mediating between the linear and the classical.

Comparison

Dynamical systems	Quantum analogue	VGF interpretation
Bounded phase space	Trace = 1; finite Hilbert	Finite stability domain
Sensitivity to initial conditions	Phase-dependent decoherence pathways	Stability–Fidelity tension
Non-separable internal structure	Entanglement	Non-factorisable loop nesting
Convergence toward	Diagonalisation	Local stabilisation

When Linear Dynamics Appears Nonlinear

The evolution of the density matrix obeys a linear equation, $\dot{\rho} = \mathcal{L}(\rho)$, yet effective nonlinearity arises when projecting, tracing out, or normalising. Conditional measurements, $\rho \rightarrow M\rho M^\dagger \,/\, \mathrm{Tr}[M\rho M^\dagger]$, introduce nonlinear feedback even though the underlying evolution remains unitary. Stochastic Schrödinger trajectories show chaotic divergence, mirroring strange-attractor behaviour in trajectory space.

Nonlinear Quantum Theories

Attempts to modify Schrödinger's equation to include explicit nonlinearity (Weinberg, Diósi–Penrose) yield superluminal signalling and contradictions with quantum statistics. Thus, physical nonlinearity must be emergent rather than fundamental.

In the VGF framework, linearity corresponds to the pre-stabilised coherence regime, while apparent nonlinearity arises from iterative local stabilisations — nested completions within the generative field.

VGF Interpretation

Level	Mathematical form	Physical / VGF meaning
Hilbert evolution	Linear, unitary $U(t) = e^{-iHt/\hbar}$	Perfect coherence; reversible iteration
Density matrix	Linear in ϱ but nonlinear under tracing	Strange attractor; coupled stabilisation loops
Classical dynamics	Fully nonlinear equations	Completed stabilisations; redundant stable loops

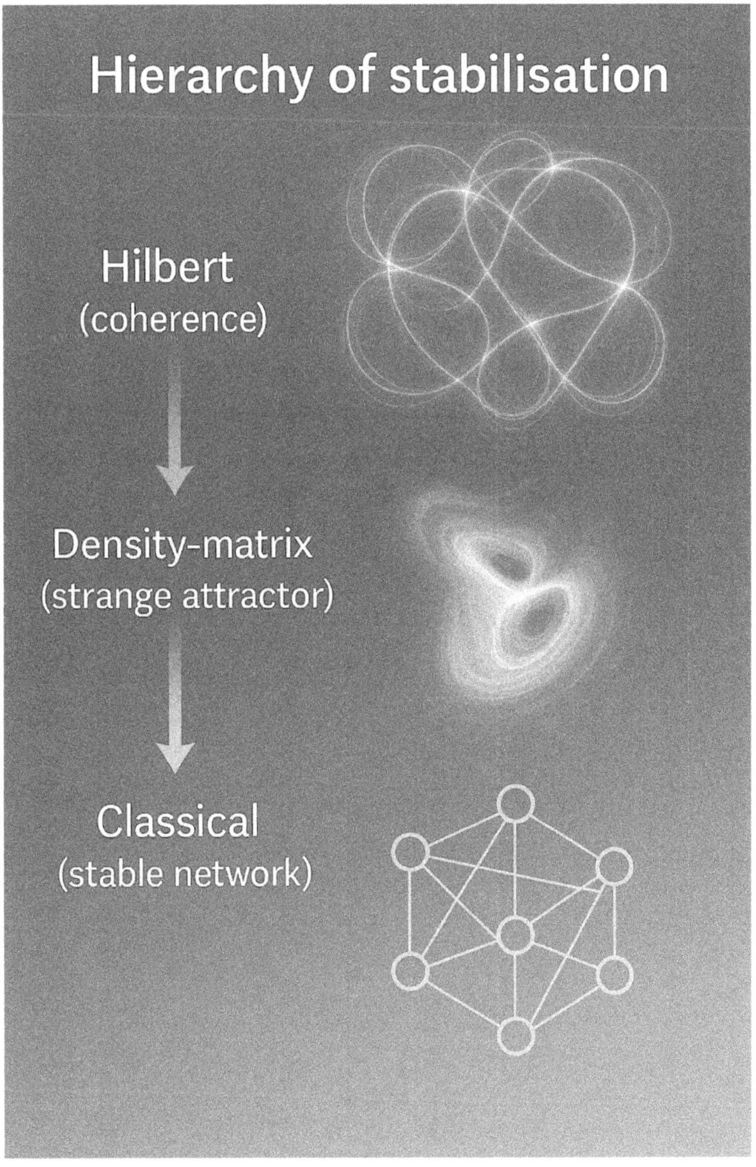

Summary

The density matrix can be seen as an intermediate attractor — a coherent network of relations whose components are locally distinct yet globally inseparable. Decoherence literally iteratively simplifies this attractor, yielding the stable pointer networks of the classical world. In the IIP-VGF framework apparent nonlinearity arises not from a modification of quantum law, but from successive local completions within the infinite iteration of the Vast Generative Field (VGF).

References

Tanimura, Y. & Kubo, R. (1989). Time evolution of a quantum system in contact with a nearly Gaussian-Markoffian noise bath. J. Phys. Soc. Jpn. 58, 101–114.

Diósi, L. (1988). Continuous quantum measurement and Ito formalism. Phys. Lett. A 129, 419–423.

Penrose, R. (1996). On gravity's role in quantum state reduction. Gen. Relativ. Gravit. 28, 581–600.

Weinberg, S. (1989). Testing quantum mechanics. Ann. Phys. 194, 336–386.

Zurek, W. H. (2003). Decoherence, einselection, and the quantum origins of the classical. Rev. Mod. Phys. 75, 715–775.

The Poincaré Recurrence Theorem

Poincaré Recurrence as the Finite Image of Infinite Iteration — The VGF Interpretation

The Poincaré Recurrence Theorem provides a classical statement of how deterministic, volume-preserving dynamics, when bounded, inevitably produce recurrence.

In the IIP–VGF framework, this recurrence is interpreted as the finite expression of infinite iteration constrained within local completion.

It represents the fidelity-saturated limit of coherence within a closed domain, just before the onset of decoherence and redundancy formation.

In this sense, recurrence is the dynamical signature of iteration under closure — infinite coherence looping within finite stability.

Classical and Quantum Recurrence

Classically, the Poincaré Recurrence Theorem states that any finite, measure-preserving dynamical system will, under infinite iteration, return arbitrarily close to its initial configuration.

Formally:

If $\Phi_t : X \to X$ is a measure-preserving flow on a finite measure space (X, μ)

then for almost every $x \in X$ and any neighbourhood U of x,

there exists a $t > 0$ such that $\Phi_t(x) \in U$.

Equivalently:

$$\liminf_{t\to\infty} \|\Phi_t(x) - x\| = 0.$$

That is, in a closed Hamiltonian system with conserved energy and bounded phase-space volume, every trajectory will eventually recur arbitrarily close to its starting point — not periodically, but infinitely often.

Quantum Recurrence

In quantum mechanics, an analogous theorem (Bocchieri & Loinger, 1957) applies to systems evolving unitarily in finite-dimensional Hilbert space.

If a state $|\psi\rangle$ evolves as

$$U(t) = e^{-iHt/\hbar},$$

then for every $\varepsilon > 0$, there exists a time $T > 0$ such that

$$\|U(T)\psi - \psi\| < \varepsilon.$$

This expresses quantum recurrence — the re-approach of a pure state to its initial condition under unitary evolution.

Both classical and quantum recurrence depend on closure, conservation, and reversibility — the absence of dissipation or environmental coupling.

VGF Interpretation

Within the IIP–VGF ontology, infinite iteration corresponds to unrestricted generativity — the limitless recursion of relations upon relations.

When this generative process is bounded within a finite relational domain — a local completion — iteration cannot expand outward.

Instead, its internal dynamics begin to loop, yielding recurrence.

Recurrence is thus iteration under closure: pure fidelity cycling within the confines of redundancy.

It represents a self-interference pattern of generativity, sustained without dissipation.

This relationship can be written symbolically as:

$$\text{IIP (infinite recursion)} \longrightarrow$$

$$\text{VGF local completion (bounded domain)} \longrightarrow$$

$$\text{Poincaré recurrence (looping coherence).}$$

When the closed system becomes open to external coupling, recurrence transitions into decoherence, loops become attractors, and fidelity condenses into stability.

Thus, recurrence represents the pre-decoherent echo of stability — fidelity circulating within a bounded relational envelope.

Morphological Comparison

System Type	Dynamical Mode	VGF Analogue	Outcome
Closed, measure-preserving	Pure unitary evolution	Infinite iteration in finite domain	Poincaré recurrence
Open, dissipative	Non-unitary evolution	Iteration coupled to environment	Decoherence; attractor formation
Expanding, hierarchical	Nested closure + re-entry	Multi-level recursion	Stability hierarchy; redundancy growth

This morphological map shows how recurrence is the boundary condition between perfect coherence and environmental openness.

At that boundary, iteration no longer generates novelty; it reproduces its own structure cyclically — a dynamic fossil of infinite recursion.

Review

Poincaré recurrence is the finite image of infinite iteration — fidelity trapped within bounded redundancy.

It represents the dynamical saturation point of coherence: infinite self-similarity looping within a closed domain.

When the domain opens to environmental coupling, recurrence dissolves into decoherence; loops become attractors, and stability emerges as redundancy.

In VGF Terms:

- Recurrence corresponds to the internal resonance of a local completion.
- Decoherence corresponds to the outward diffusion of that resonance into redundant copies.
- The stability–fidelity law governs the transition between the two.

Thus, Poincaré recurrence is the dynamical analogue of idempotence in the IIP field:

$$\Phi_t(E) = E,$$

reached not by instantaneous closure, but rather by the infinite re-approximation of fidelity to itself.

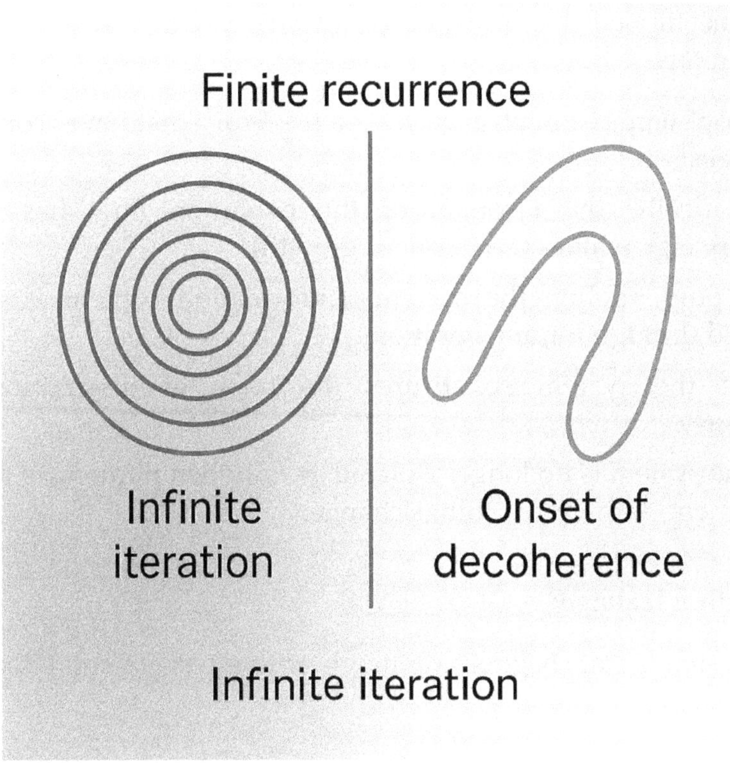

Figure: From Infinite Iteration to Finite Recurrence: Bounded Coherence and the Onset of Decoherence.

Left: dense phase-space loops transitioning into

Right: open attractors (right).

The left region represents pure, closed-system fidelity; the right, the beginning of stabilised redundancy.

Loops Become Attractors

In the closed system: recurrence = looping coherence.

In a Poincaré-recurring system:

- The system is measure-preserving (no dissipation).
- Every point in phase space evolves reversibly.
- Orbits are quasi-periodic — they fill out tori or closed contours.

- No trajectory ever settles; it keeps revisiting near its start state indefinitely.

So, a "loop" here is a neutral cycle — a trajectory that never converges or diverges.

Mathematically, the Jacobian of the evolution flow has purely imaginary eigenvalues (no decay or growth).

This is pure fidelity: the system's information content is exactly preserved through infinite iteration.

Opening the system constitutes the birth of dissipation and redundancy

When the system is no longer isolated — *i.e.* when it interacts with an environment — three key things change:

Information leakage:

- The system's state begins to entangle with environmental degrees of freedom.

Phase dispersion:

- Different parts of the loop begin to lose phase coherence because each interacts differently with the environment.

Volume contraction:

- The flow is no longer measure-preserving — it becomes dissipative. Liouville's theorem (which guarantees recurrence) no longer holds.

The phase-space trajectory that used to be a neutral loop now contracts toward a stable manifold.

Instead of endlessly revisiting its past states, it now converges to a region — an attractor.

Dynamical Picture:

In equations of motion (say for a simple oscillator with damping):

$$\dot{x} = v,$$

$$\dot{v} = -\omega^2 x - \gamma v.$$

- When $\gamma = 0$: the motion is perfectly periodic — recurrence loop.
- When $\gamma > 0$: the motion spirals inward — it converges to an attractor (a fixed point or limit cycle).

So:

$$\text{Loop}\ (\gamma = 0) \ \Rightarrow \ \text{Attractor}\ (\gamma > 0).$$

This is the mathematical transition from recurrence to decoherence.

The "loop" becomes an "attractor" once dissipation or openness is introduced.

In Quantum–Decoherence Terms

A closed, unitary quantum system has purely oscillatory dynamics:

$$\rho(t) = U(t)\rho(0)U^{\dagger}(t),$$

and this evolution is reversible → quantum recurrence applies.

Once we trace out the environment:

$$\rho_S(t) = \mathrm{Tr}_E[U(t)\rho(0)U^{\dagger}(t)],$$

we introduce an effective damping of the off-diagonal elements (the coherences).

The state no longer loops perfectly through Hilbert space — it relaxes toward a diagonal ensemble (the pointer basis).

That diagonal ensemble is the attractor in the space of density matrices — the state to which the system converges under environmental interaction.

In IIP–VGF language:

Closed (pure iteration)	Open (decohering iteration)
Infinite fidelity within finite loop	Fidelity converts to redundancy
No loss of relational amplitude	Relations stabilise through repetition
Self-consistent recurrence (E = I(E))	Self-consistent idempotence ($E^2 = E$)
Coherence "rehearses" itself endlessly	Coherence "solidifies" into stable attractors

So in the VGF framework:

- The recurrence loop is the pre-stable phase — a pure iteration that has not yet committed to a fixed relation.
- The attractor is that same loop after it has folded into itself through redundancy — when its outward iterations have been absorbed into stable, redundant self-consistency.

Formally, we could think of this as:

$$\text{Iterative recurrence: } \mathscr{K}n + 1 = f(\mathscr{K}n) \text{ becoming}$$

$$\text{Idempotent attractor: } \mathscr{K}* = f(\mathscr{K}*).$$

In other words, recurrence is the iterative phase of approach to idempotence.

Review

In a closed system, recurrence manifests as fidelity looping through finite relational space: coherence without loss.

When the system opens to the environment, those loops become dissipative — their trajectories contract toward stable attractors.

The recurrence loop thus becomes the germ of an attractor: a rehearsal of stability that, once coupled to redundancy, collapses into a fixed relational form.

Symmetry and Supersymmetry

Symmetry in the VGF

In the VGF conception, symmetry is not imposed upon an already-formed structure but arises as the first act of self-agreement within infinite iteration.

A symmetry is the minimal condition under which an iteration can repeat without contradiction — the state in which a relation becomes identical with its own reproduction. This primordial self-coincidence marks the birth of stability: the first fixed-point in the vast generative flow.

From such self-agreement, more elaborate invariances unfold. Spacetime, rotational, and gauge symmetries represent progressively differentiated projections of this primal coherence, each expressing a way in which the field preserves identity through transformation.

In this sense, symmetry is not a property of things in space or time; it is the rule that allows spacetime-like order to appear at all — the grammar of self-recognition by which the generative field stabilises its own activity.

Noether's theorem, which in physics links every continuous symmetry to a conserved quantity, can then be read as a formal shadow of the Stability–Fidelity Law.

Each conservation law preserves the memory of a symmetry through the decohering environment, ensuring that certain invariants — energy, momentum, charge, spin — remain stable even as the finer structure of iteration is lost. The price of that stability is fidelity: the original generative richness thins into discrete quantum numbers and measurable constants. Thus, what we call a "conserved quantity" is the fossil of an iterative self-agreement — a record of how the field once repeated itself without contradiction. Symmetry, in this view, is both the origin and the enduring trace of the generative process: the law by which iteration recognises itself across transformation, and through which the physical world retains coherence amidst continual change.

Symmetry as Idempotent Iteration

In the VGF framework, a symmetry is an idempotent IIP: an iteration that re-applies to itself without change. Mathematically, it is a morphism or operator f satisfying $f \circ f = f$, defining an invariant subspace of the generative field. In physical language, this corresponds to a projector onto a symmetry sector — the state-space that remains unchanged under the relevant transformation group. Every conservation law, therefore, expresses an idempotent relation within the generative field: a stabilised loop that reiterates itself perfectly. Symmetry breaking arises when such idempotence is perturbed, allowing the field to regain generative differentiation. Thus, the universal symmetries of physics are the idempotent stabilisations of iteration itself — fixed points of the infinite self-action that underlie all measurable invariants.

$$f \circ f = f$$

$$f : \Phi \to \Phi$$

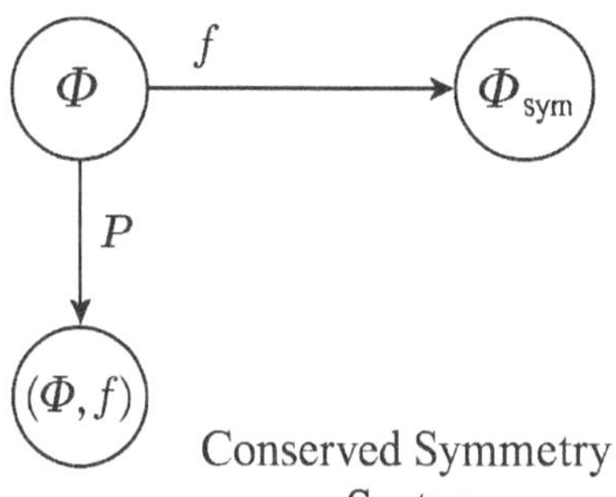

Conserved Symmetry Sector

Supersymmetry as Meta-Idempotent Iteration

In the VGF framework, supersymmetry represents an attempted unification of two stabilisation modes — fermionic (object-forming) and bosonic (relation-mediating) — under a higher-order idempotent iteration.

Whereas ordinary symmetries satisfy $f \circ f = f$, supersymmetry satisfies $Q^2 = P_\mu$: its self-iteration reproduces translation rather than identity, linking internal generative structure to spacetime propagation. The experimental absence of superpartners can then be viewed not as a failure but as evidence of decohered meta-idempotence — a broken equilibrium between dual stabilisation regimes. In this reading, SUSY is the echo of a more general generative law in which perfect boson–fermion reciprocity is lost through redundancy, yielding the asymmetric stability of the observable universe.

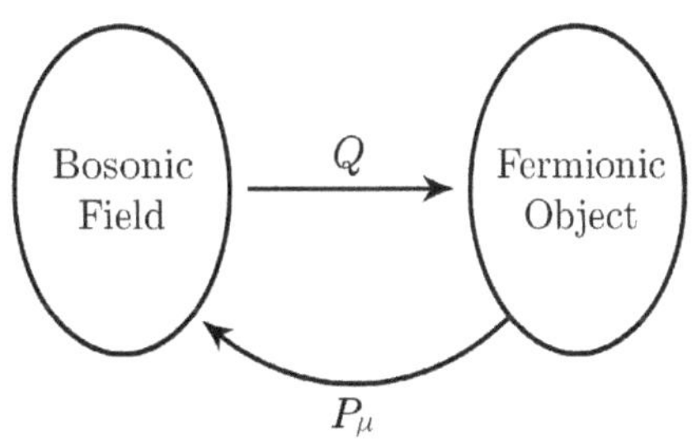

The Measurement Problem

In the IIP–VGF approach, the density matrix is not an adjacency matrix between a network of pre-existing diagonal nodes. Rather, it is itself a network of relational primitives, each element of which is a node in its own right.

This means:

- Each matrix element ρ_{ij} (whether diagonal or off-diagonal) corresponds to a relational micro-object — an IIP stabilisation of the correlation between basis components i and j.
- Even if $\rho_{ij} = \rho_{kl}$ numerically, the relational context (its placement in the global web) differentiates them.
- The matrix as a whole, therefore, represents a field of coupled micro-loops — a small but self-contained VGF fragment.

So: ρ is not a table of relations between states, but rather, a local relational field whose topology changes under decoherence.

Partitioning and Tracing Out as Relational Contraction

When we partition into system and environment and trace out the environment E, we are effectively collapsing a two-layer network (ρ_{SE}) into a single effective network (ρ_S).

In VGF Terms:

- The operation "trace out" is a relational contraction — the absorption of one part of the web into the redundancy of the rest.

- The surviving structure (the diagonals) are the self-referential loops that remain coherent after this contraction.

Arranging those surviving loops as a ring is natural — it reflects the cyclic symmetry of the pointer ensemble:

- Each diagonal element is a node on the ring.
- The faint residual coherences (the near-zero off-diagonals) are the faded chords across the diameter.

The Next Decoherence: Measurement as Further Contraction

In the IIP-VGF approach then, measurement constitutes another iteration of decoherence. In IIP terms, every measurement event is another round of environmental completion — another local stabilisation of redundancy.

That means:

- The ring of potential outcomes is not yet classically closed.
- Each node on that ring is itself an open link capable of further coupling.
- A "measurement" occurs when only one of those nodes forms a stable link outward into a wider web — the surrounding environment of other rings.

This gives us the second-order decoherence:

$$\text{Ring of candidate loops} \longrightarrow \text{One loop linking outward to next ring}\,.$$

That is, the measurement is the selection of the next relational continuation.

The Chain of Rings: A Topology of Natural Determinism

When each ring (a decohered ensemble of possibilities) passes on only one connection to its successor, the universe's web of realised outcomes becomes a chain, not a mesh:

- Each "ring" = a moment of decohered redundancy (a set of alternatives).
- Each "link" = the one realised coupling that continues into the next completion.
- The chain as a whole = the unfolding classical worldline, or the deterministic network of measurables.

In categorical language, this is a monoidal chain of idempotents:

$$R_1 \xrightarrow{f_1} R_2 \xrightarrow{f_2} R_3 \xrightarrow{f_3} \dots$$

where each R_n is the "ring" of redundant pointer states, and each morphism f_n is the unique surviving coupling.

This explains why classical laws look deterministic:

- Within any given segment, the redundancy is vast (stability).
- Across segments, only one continuity is realised (causality).
- The overall structure is a chain of nested decoherences.

Why Loops Do Not Close

The "rings" form a chain, each "ring" connecting to it "environment" by just one node. Thus, the chain cannot form loops.

If two rings were to connect via more than one node, this would create coherence revival — feedback of phase information.

That would correspond to non-Markovianity, quantum revivals, or closed timelike correlations.

So the absence of such loops is what defines the deterministic classical limit of quantum decoherence:

Classical Causality =

Decoherence Chain with no Re-entanglement Cycles.

In Other Words:

- Quantum domain → dense network, multiple feedback loops.
- Classical domain → one-path chain of rings (tree-like structure).

Thus, the loss of coherence is actually a topological pruning rule in the VGF.

Unifications

- Density-matrix mathematics (as contraction of correlations).
- Quantum Darwinism (environmental selection of one pointer path).
- Category-theoretic iteration (composition of idempotent morphisms).
- Stability–Fidelity Law (redundancy growth *vs.* fidelity loss).

In the IIP–VGF picture, the chain of rings is simply the classical causal skeleton — the remnant of infinite relational iteration pruned by successive local completions.

Review

I. At the VGF level, the density matrix represents a web of relational primitives — not relations *between* distinct objects, but relations that *are* objects.

II. Tracing out the environment contracts this web into a ring of stable loops — the pointer basis.

III. A further iteration of decoherence (measurement) selects one node of the ring for outward coupling to another ring.

IV. The world of classical measurables thus forms a chain of rings — each a local completion — linked by unique couplings.

V. Loops between rings do not form: they would re-introduce coherence.

VI. Causality, therefore, is the topology of nested decoherences — the linear unfolding of the VGF's infinite recursion.

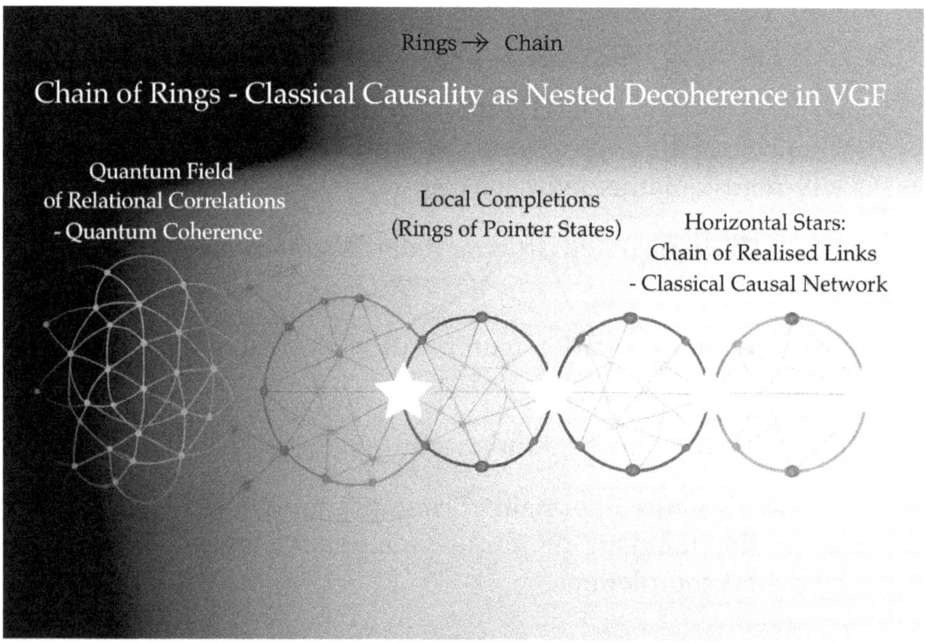

Zero Point Energy

Introduction

In popular accounts of quantum field theory (QFT), the vacuum is often described as a sea of fluctuating fields where virtual particles "pop in and out of existence". While evocative, this image misrepresents what actually occurs.

The formal structure of QFT does not describe tiny particles materialising and vanishing, but rather a continuous field whose correlations fluctuate according to the uncertainty principle. The vacuum is the ground state of the field — not empty space, but a state of maximum coherence and minimum redundancy.

Zero-Point Energy

Each mode of a quantum field behaves like a harmonic oscillator with a lowest energy:

$$E_0 = \frac{1}{2}\hbar\omega .$$

Even when no excitations are present, the uncertainty principle forbids the field from being exactly at rest, so every mode contributes a small zero-point energy. Summing these contributions leads to a formally divergent vacuum energy density — the 'vacuum catastrophe' — unless carefully regularised:

$$\rho_{\text{vac}} = \sum_{\text{modes}} \frac{1}{2}\hbar\omega \to \infty .$$

Even with a high-frequency cutoff (say, the Planck scale), the energy density is still 10^{120} times too large compared with the observed cosmological constant — this is the vacuum catastrophe, so-called because it appeared to be a catastrophe for scientific prediction.

In physical terms, these fluctuations are not literal movements of particles but potentialities of the field's state-space. They represent correlations that can, under suitable conditions, decohere into measurable excitations — photons, electrons, phonons, etc. Thus, 'virtual particles' are bookkeeping devices in perturbative expansions, not transient lumps of matter appearing and disappearing.

Decoherence and Measurement

Decoherence provides the link between the coherent vacuum and the measurable classical world. When a field interacts with an environment, phase relations between its components are lost, and certain configurations stabilise as 'pointer states.' What we call 'particles' are those decohered excitations — stable relational attractors in the generative field. Measurement does not create them *ex nihilo*; it stabilises one of the many relational possibilities.

- The vacuum corresponds to the highest-fidelity level of the VGF—pure iteration with no stabilised redundancies.
- The so-called zero-point energy is the latent measure of iteration amplitude across all possible relational frequencies, before any decoherence has created stable energy loops.
- In IIP-VGF terms a "particle with energy E = hν" is the first stabilised attractor in which that latent iteration condenses into redundant recurrence—a closed loop in phase.

Thus, in IIP-VGF terms the actual energy doesn't exist before decoherence—it emerges as one of the stable attractor parameters when the VGF collapses locally into measurable form.

So the "zero-point energy" is not a hidden variable or pre-existing reservoir of energy. Rather, it's a mathematical fingerprint of the field's coherent relational structure, misread through the lens of classical energetics.

Statement	Status
The vacuum "contains energy"	Metaphorical — it contains correlation amplitudes.
Energy is definite before measurement	False (unless you assume hidden variables).
Zero-point energy = decohered observable	No — it's an expectation value in the absence of decoherence.
"Popping particles"	Narrative shorthand for momentary correlation amplitudes in perturbation theory.
In IIP–VGF terms	A representation of latent iteration potential before stabilisation (pre-energetic coherence).

IIP–VGF Interpretation

Within the IIP–VGF framework, the vacuum corresponds to the most coherent, least stabilised level of the Vast Generative Field. Zero-point energy represents the latent, pre-decoherent iteration activity within that field. Decoherence progressively transforms coherence into redundancy — forming stable attractors. At the limit, the classical world emerges as a dense network of redundant relations.

Mapping Quantum Field Terms to IIP–VGF Concepts

Quantum Field Concept	IIP–VGF Interpretation
Vacuum field	Fully coherent region of the VGF (undifferentiated iteration space)
Zero-point fluctuations	Latent internal iterations within the field
Virtual particles	Transient self-relational corrections (not objects)
Decoherence	Partitioning of VGF coherence into stable relational attractors
Measurement	Stabilisation of a relational attractor (pointer-state fixation)

Why Zero-Point Energy Is Not Gravitationally Real: The IIP–VGF Resolution of the Vacuum Catastrophe

By comparing the conventional quantum field view with the frameworks of decoherence and the Vast Generative Field (IIP–VGF), it becomes clear that the apparent vacuum energy is not a physical reservoir but a measure of undecohered coherence. The discrepancy known as the 'vacuum catastrophe' arises from confusing fidelity (pre-decoherent iteration potential) with stability (post-decoherent redundancy).

The Quantum–Gravitational Mismatch

In general relativity (GR), any energy density should curve spacetime according to Einstein's equation:

$$G_{\mu\nu} + g_{\mu\nu}\Lambda = \frac{8\pi G}{c^4}T_{\mu\nu}$$

If the vacuum energy density calculated by quantum field theory (QFT) were gravitationally real, it would lead to catastrophic curvature of spacetime, incompatible with the observed large-scale structure of the universe. The predicted value exceeds the observed cosmological constant by about 120 orders of magnitude.

In QFT, however, the zero-point term is a constant offset in the Hamiltonian that does not affect measurable energy differences or interactions. Only differences matter in quantum mechanics, not absolute scale. Thus, the zero-point energy represents a formal expectation value, not a physical reservoir.

Decoherence View

From the perspective of decoherence theory, energy becomes real only when correlations between system and environment stabilise into redundant, classical records. The zero-point term belongs to the undecohered part of the field's correlation structure—it never stabilises as a measurable quantity. Hence, it cannot produce gravitational curvature because gravity, as a classical effect, responds only to redundant (stabilised) relations.

Only decohered, redundant energy distributions are gravitationally real. The 'zero-point energy' of QFT exists as an amplitude of coherence, not a source of curvature.

IIP–VGF Interpretation

In the IIP–VGF framework, the vacuum corresponds to the fully coherent, fidelity-rich level of the Vast Generative Field. Zero-point 'energy' represents the pure iteration potential of the VGF—relations that have not yet decohered into redundant, stable loops. Gravitation arises only from stabilised, redundant structures that define classical spacetime curvature.

Thus, zero-point energy, though large in mathematical expectation, has no gravitational reality because it belongs to the domain of pure fidelity, not stability. The cosmological constant problem therefore reflects a category error: mistaking the fidelity of the coherent field for the redundancy of stabilised matter–energy.

Conceptual Synthesis

- Zero-point energy: formal expectation of field coherence, not a measurable energy reservoir.
- Gravitation: curvature produced by redundant, stabilised relations —the classical decoherence end-state of the VGF.
- The mismatch: arises from projecting pre-decoherent fidelity (QFT) into the post-decoherent regime (GR).

In essence, gravity couples to stability, not potentiality.

Interpretive Comparison: Zero-Point Energy Across Frameworks

Concept	Quantum Field Theory (Conventional)	Decoherence Interpretation	IIP–VGF Interpretation
Nature of the vacuum	A physical field with non-zero energy in every mode	A global superposition of correlations with no stabilised redundancy	Pure fidelity domain of the VGF: pre-stabilised iteration potential
Zero-point energy	Sum of (½ħω_k) over modes; formally infinite before renormalisation	Expectation value of undecohered field correlations	Latent generative coherence; not energy but iteration amplitude
Why it doesn't gravitate	Constant shift in Hamiltonian; only differences matter	Gravity couples only to decohered, stable correlations	Gravitation emerges only from redundant stabilisations
Cosmological constant problem	$\rho_v a c^Q F T \gg \rho_o b s$	Miscounting coherence as observable	Category error: treating fidelity as stability
Ontological status	Real but hidden physical energy	Non-redundant coherence amplitude	Pure iteration potential within infinite IIP network

Review

In QFT, the vacuum is an energetic field; in decoherence, it is a correlation structure; in the IIP–VGF framework, it is an infinite iteration domain. Gravitation belongs to the realm of stabilised redundancy. Zero-point energy belongs to the realm of pure coherence. Their mismatch—the vacuum catastrophe—is not a paradox of physics, but a misalignment between domains of fidelity and stability.

Vacuum Effects

A Central Controversy:

View	Claim	Consequence
Operational / decoherence view	Energy is a measurable quantity only when a system decoheres with a measuring apparatus. The ground-state expectation value is not a real, possessed energy—just a property of the probability amplitude.	Zero-point energy is not physically real; it's a formal offset.
Field-reality view (QFT mainstream)	The field Hamiltonian's ground-state expectation value contributes to observable effects such as the Casimir force, Lamb shift, and spontaneous emission.	Vacuum fluctuations are physically real even without measurement.

Conventional Interpretation (Quantum Field Theory)

In standard QFT, the vacuum is not empty but contains zero-point fluctuations. Systems interacting with this vacuum experience measurable shifts or forces arising from those fluctuations:

Effect	Description in QFT
Casimir Effect	Attractive force between plates due to altered zero-point mode density between boundaries.
Lamb Shift	Energy-level correction in hydrogen due to the electron's coupling with vacuum fluctuations.
Spontaneous Emission	Emission of a photon from an excited atom 'stimulated' by vacuum field modes.
Vacuum Energy	Ground-state expectation value of the field Hamiltonian, producing virtual particle effects.

IIP–VGF Interpretation (Decoherence and Stability)

In the IIP–VGF framework, the vacuum corresponds to the most coherent level of the generative field — a domain of iterative relations without stabilised redundancy. Observable effects arise when boundary conditions or interactions force local decoherence, partitioning the VGF into stable attractors and measurable events:

Effect	Traditional interpretation	IIP–VGF reading
Casimir force	Suppressed vacuum modes between plates	Local partition in the VGF — environmental constraint forming a partial decoherence. Reduced relational redundancy due to the imposed boundary partition
Lamb shift	Energy correction from vacuum fluctuations	Micro-adjustment of local attractor structure in relational field
Spontaneous emission	Vacuum field stimulates decay	Decoherence-induced release of a stable iteration (photon loop)
"Vacuum is physically real"	Field filled with virtual particles	Latent iteration domain; stability arises only through interaction

Synthesis

Both frameworks describe the same measurable outcomes. The distinction lies in what is taken as fundamental: in QFT, fluctuations are causes acting on matter; in IIP–VGF, they are relational possibilities that become manifest only through decoherence. Thus, the so-called 'vacuum energy' is better understood as the fidelity-rich coherence of the generative field, whose partitioning gives rise to stability, measurement, and observable energy exchange.

Fractal Hierarchies

In the IIP-VGF framework treatment of Hilbert space, idempotents ($P^2 = P$) correspond to stabilised distinctions or projectors. They act as IIP attractors: once applied, further iteration yields no change.

A key open question is how the infinite, fractal hierarchy of such idempotents organises into effective, finite-dimensional sectors that characterise the classical and quantum objects we observe.

Idempotents as Stability Atoms

- Projectors $P^2 = P$ are the canonical idempotents.
- They stabilise discriminations in Hilbert space.
- In the VGF picture: idempotents = IIP attractors, representing fixed points of infinite iteration.
- Building a Fractal Hierarchy

Three processes generate hierarchies of (approximate) projectors:

- Decoherence: suppresses off-diagonals in a pointer basis, producing nested projectors.
- Coarse-graining: finite resolution of instruments $\rightarrow$ spectral bands as projectors onto subspaces.
- Redundancy (Darwinism): intersecting environment records stabilises further projectors.
  Together these create a lattice of projectors, refined self-similarly: a fractal hierarchy.

From Infinite Hierarchy to Finite Sectors

- Real environments/instruments produce approximate projectors (effects) that stabilise finite subspaces.
- Repeated decoherence drives states toward block-diagonal form; off-block coherences fall below noise floors.
- The centre of the effective operator algebra (commuting idempotents) defines superselection sectors.
- Each block behaves like a finite-dimensional algebra $M_d(C)$, *e.g.* qubits, qdits, or classical registers.
- Renormalisation viewpoint: coarse-graining prunes the fractal hierarchy into AF (approximately finite) algebras, represented by Bratteli diagrams.

VGF / IIP Translation

- IIP attractors: idempotents are infinite-iteration fixed points.
- Fractal layering: iterate discriminate → stabilise → refine = self-similar tower of projectors.
- Finite sectors: truncating at finite fidelity/stability budget yields finite-dimensional blocks.
- These blocks are the effective 'objects' of physics.

Significance

- This explains why finite-dimensional models (qubits, qdits) capture physical systems.
- It clarifies the emergence of classicality: classical variables = central idempotents whose refinement has saturated.
- Connects decoherence/Darwinism to operator-algebra structures: projector lattices, AF algebras, Bratteli diagrams.

Thus the fractal hierarchy of idempotents does not remain infinite at the operational level: decoherence, redundancy, and coarse-graining organise it into finite-dimensional effective sectors, which appear as the qubits, particles, and classical observables of our world.

The Big Bang in the IIP-VGF Framework

Planck Epoch ($t < 10^{-43}$ s) — Pure iteration

At the boundary of the Planck Epoch all known physical quantities lose independent definition — spacetime, mass, and even causality dissolve into quantum–gravitational indeterminacy.

In IIP–VGF terms this corresponds to the pure IIP phase:

- Iteration without stable self-reference.
- No consistent boundary between "system" and "environment".
- Infinite potential for self-relation, but zero stabilisation.

This regime is pre-symmetry, pre-metric, and pre-object.

It is the Vast Generative Field in its raw state, before any idempotent loops have condensed.

Inflation and Early Quantum Field Epochs ($10^{-43} \rightarrow 10^{-32}$ s)

Here, quantum fluctuations become magnified into macroscopic structures.

Symmetry breaking occurs:

- The strong and electroweak forces separate.
- The Higgs field stabilises.
- Particles acquire mass.

In IIP–VGF language, this is when proto-objects begin to form: redundant self-consistent loops (absorptive idempotents) that define distinct stabilisation channels.

It is not yet classical, but it is the first appearance of iterative redundancy that can support stability — the generative "seed" of classical law.

Reheating and Baryogenesis ($\sim 10^{-32} \rightarrow 10^{-6}$ s)

Matter–antimatter asymmetry locks in; quarks bind into hadrons; decoherence becomes directional.

Here, energy-density gradients and symmetry breakings introduce persistent asymmetries — necessary for lawlike behaviour.

This is the decoherence epoch in cosmological terms: the point where iteration begins to prefer certain attractor routes — *i.e.* the proto-Stability–Fidelity Law in operation.

Recombination and the CMB (~380,000 years)

Electrons and nuclei combine to form neutral atoms; photons decouple and travel freely.

The CMB is the first global signature of a fully decohered field — a redundancy pattern that retains memory of quantum origins but behaves classically.

In IIP–VGF terms, this is the global stabilisation of the generative field: the moment when redundancy has grown so extensive that local fluctuations can persist as classical structures.

The field now supports objectivity: systems can maintain identity across scales of observation.

Thus, this is the establishment of classical law in the cosmological sense.

Reionisation and the first stars (~180 Myr)

The emergence of stars marks a secondary iteration: gravity amplifies classicality by creating hierarchical, nested attractors.

This corresponds to the second-order stabilisation of the IIP–VGF — the beginning of nested decoherence and the layered causal order we associate with the modern classical universe.

Summary in IIP–VGF Terms

Epoch	IIP–VGF description	Status relative to classicality
Planck ($\leq 10^{-43}$ s)	Pure iteration, no redundancy	Pre-law
Inflation / symmetry breaking (10^{-43}–10^{-32} s)	First idempotent stabilisations	Proto-law
Reheating / baryogenesis	Directional decoherence	Law formation
Recombination / CMB (3.8×10^{5} yr)	Global redundancy field, first objective structure	Classical law established
Reionisation / first stars (1.8×10^{8} yr)	Nested stabilisations, hierarchy of classical attractors	Expansion of classical regime

Review

In the IIP–VGF framework , the classical laws of physics emerge not from the absolute beginning but from the progressive stabilisation of iteration through redundancy. The Planck and inflationary epochs correspond to pre-stabilised iteration; recombination marks the first universe-wide decoherence where redundant self-relations form a stable, law-bearing field.

The cosmic microwave background is therefore the signature of the moment when the generative field became sufficiently redundant to support classical objectivity — the crystallisation of law from iteration.

The CMB as the Limit of Observability

In IIP–VGF terms:

Recombination represents the transition from an internally coupled generative regime to an externally coherent observational regime:

- Before recombination: the Universe is a fully entangled field, with no stable separation between emitter, medium, and observer — pure IIP iteration without redundancy.
- At recombination: the field decoheres enough to create distinct, stable relations — photons that travel independently, matter that retains identity, and redundancy that makes observation possible.
- Beyond recombination: objectivity emerges — the ability to re-identify a structure across time and distance.

Thus, in VGF terms:

The CMB marks the outermost visible boundary of the field's first great decoherence — the moment the generative iteration achieved enough redundancy to make its stabilisations observable.

In this sense, we cannot "see beyond" recombination not because something is physically blocking us, but because before that epoch there was no stable difference between emitter, medium, and signal — i.e. no classical information channel existed yet.

Spacetime as an Emergent Stabilisation

Epoch	Physical picture	IIP–VGF interpretation
Pre-Planck ($t < 10^{-43}$ s)	Quantum-gravitational foam, no smooth geometry	Pure infinite iteration — relations without metric or ordering; no distinction between "before" and "after".

Planck (~10^{-43} s)	Quantum gravity scale	First stabilised *ordering relations* appear — the proto-notion of succession (emergent "time") arises as the field differentiates its own internal iteration.
Post-Planck → Inflation (10^{-43}–10^{-32} s)	Rapid expansion, vacuum symmetry breaking	Redundant relational patterns between regions become measurable — proto-spatial differentiations emerge; the coherence between succession (temporal order) and extension (spatial relation) yields the spacetime manifold.
After inflation	Smooth metric spacetime	Fully decohered spacetime: classical geometry as a global redundancy pattern.

In the logic of the IIP-VGF time is expected to appear first:

In pure iteration, the only primitive is repetition:

$$\odot \Rightarrow \circledcirc \Rightarrow \circledcirc \Rightarrow \dots$$

That repetition already implies succession, even before any metric or geometric structure exists.

So proto-time — the ordered self-application of the generative operation — is the first coherence to form.

Only when different iterative lines begin to co-stabilise relative to one another does space emerge as relational differentiation.

Thus:

Time arises from iteration; space arises from relational consistency among simultaneous iterations; spacetime appears when both stabilise jointly.

Resolution of the "From Zero" Paradox:

- At zero, there is no space and no stable time — only unstructured iteration (potential).
- Around the Planck epoch, iteration stabilises enough to generate temporal ordering.

- Slightly later, redundancy among distinct stabilisations produces spatial extension.
- Their joint decoherence yields spacetime as a composite invariant.

Review

In the IIP–VGF framework, spacetime does not pre-exist the generative field but arises from it through successive stabilisations.

The first emergence is temporal: repetition itself defines ordered iteration.

Spatial relations appear when redundant iterations stabilise relative to one another, giving extension and locality.

The coupling of these two invariances — order and relation — produces spacetime as a stable manifold.

Thus, "time" appears as the field's first act of self-reference; "space" as its first act of mutual reference; and "spacetime" as their coherent redundancy.

Classical spacetime therefore appears around the Planck epoch, rather than existing fully formed at zero.

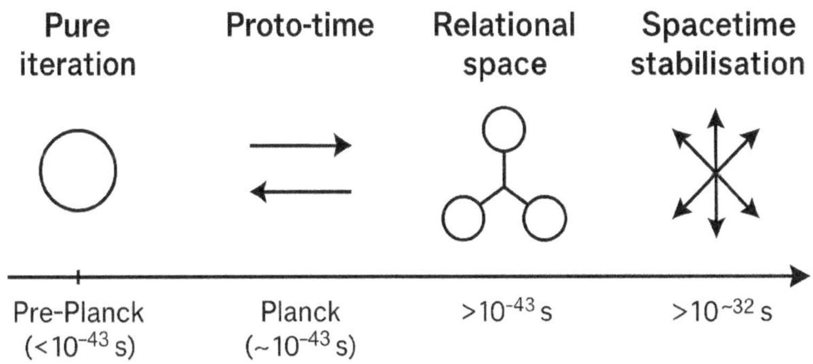

In the IIP–VGF framework, *proto-time* emerges first from ordering relations, relational space then emerges from spatial extension, and spacetime is stabilised when both invariances cohere.

General Relativity - The VGF Field Equation

Preface: Origin of the VGF Field Equation

General Relativity tells us how curvature and energy correspond. The VGF framework tells us **why** they correspond — because both are projections of the same generative law operating at different levels of decoherence.

The following formulation extends Einstein's curvature–energy relation to a more general symmetry between local stabilisation and global coherence. It originates in the Infinite Iteration Principle (IIP) and the Vast Generative Field (VGF) conception, in which all physical laws arise as fixed-point balances within a continuously self-iterating field. The tensors $\mathscr{S}\mu\nu$ and $\mathscr{F}\mu\nu$ represent the complementary aspects of this balance: $\mathscr{S}\mu\nu$ encodes local stabilisation (self-sustaining iteration, or "stability"), while $\mathscr{F}\mu\nu$ encodes global coherence (redundant consistency, or "fidelity"). Their equilibrium,

$$\mathscr{F}\mu\nu = \kappa\,\mathscr{S}\mu\nu,$$

reduces to general relativity when the two aspects decouple ($D_{\mu\nu} = C_{\mu\nu} = 0$) and extends it when their mutual feedback becomes significant. The theory therefore interprets spacetime curvature and energy–momentum not as separate entities but as complementary stabilisations of the same generative process.

Interpretive Overview: The Fidelity–Stability Field

The VGF Field Equation $\mathscr{F}\mu\nu = \kappa\, \mathscr{S}\mu\nu$ expresses the Stability–Fidelity Law in geometric form. It replaces the separation between matter and geometry with a single iterative balance inside the Vast Generative Field.

1. Meaning of the Two Tensors

Symbol	Role in physical language	Meaning in IIP–VGF terms
$\mathscr{S}\mu\nu$	Matter / energy / local dynamics	Stability tensor: the record of self-sustaining iterations — all loops that maintain internal identity.
$\mathscr{F}\mu\nu$	Geometry / curvature / global coherence	Fidelity tensor: the record of field redundancy — how coherence is preserved across the whole field.

Local stabilisations (mass, energy, information storage) create curvature, and curvature in turn constrains further stabilisation. Neither side exists without the other.

2. The Balance Principle

Differentiating the field equation gives the exchange law:

$$\nabla^{\mu} D_{\{\mu\nu\}} = \kappa\ \nabla^{\mu} C_{\{\mu\nu\}}$$

This states that every increment of local order (stability) must be accompanied by an equal increment of global coherence (fidelity). The field maintains self-consistency: energy is never isolated from geometry; redundancy is never detached from structure. Classical GR corresponds to the limit where this exchange vanishes (D=C=0).

3. Physical interpretation

Phenomenon	Description in this framework
Mass–energy	Concentrated self-iteration: stability locked into local loops.
Gravity / curvature	Redundant coherence of those loops: the field's global compensating
Quantum coherence	Fine-grained stabilisation before redundancy dominates.
Decoherence / classicality	Growth of redundancy $D_{\mu\nu}$ — fidelity spreads, stability thins.
Dark energy / inflation	Large-scale redundancy pressure driving expansion.

The familiar Einstein relationship between curvature and energy is therefore the low-redundancy limit of a deeper generative symmetry.

4. Intuitive Picture

1. Iteration begins → pure potential (no distinction between system and field).
2. Self-consistent loops form → stability appears (positive energy).
3. Field coherence adjusts → fidelity grows (gravitational curvature, negative potential).
4. Equilibrium achieved → $\mathcal{F}\mu\nu = \kappa\,\mathcal{S}\mu\nu$.

The universe at every scale is this continual feedback: stability concentrating, fidelity compensating. Space and matter are the two halves of the same self-recognition.

5. Emergent Interpretation of Einstein's Equation

Einsteinian term	VGF translation
Curvature $G_{\mu\nu}$	Fidelity: coherence of the generative field.
Stress–energy $T_{\mu\nu}$	Stability: self-iteration and information retention.
Equality sign =	Iterative equilibrium — the field recognising itself across scale.

Hence: Geometry ↔ Matter ⇒ Fidelity ↔ Stability.

6. Cosmological significance

At cosmic scales, this balance becomes dynamic:

$$\dot{X}\rho + 3H(\rho + p) = +Q, \quad \dot{X}\rho_C + 3H(\rho_C + p_C) = -Q$$

with Q the rate of exchange between local structure and global coherence. Inflation, dark energy, and the quantum-to-classical transition all represent different regimes of this same exchange.

Summary

The VGF Field Equation turns Einstein's curvature–energy reciprocity into a universal generative principle. Spacetime geometry and energy density are no longer independent quantities but complementary stabilisations of the same iterative field. Stability creates curvature; curvature sustains stability — a perfect generative balance.

The VGF Field Equation: the Fidelity–Stability Balance

In the VGF framework, spacetime geometry and matter–energy are dual aspects of a single generative equilibrium. The Einstein relation

$$G_{\mu\nu} = \frac{8\pi G}{c^4} T_{\mu\nu}$$

is generalised to:

$$\mathscr{F}\mu\nu = \kappa\,\mathscr{S}\mu\nu$$

Here $\mathscr{F}\mu\nu$ is the Fidelity tensor (redundant coherence / curvature), $\mathscr{S}\mu\nu$ is the Stability tensor (localised iteration / energy–momentum), and κ reduces to $8\pi G/c^4$ in the GR limit.

Constitutive Structure

$$\mathscr{F}_{\mu\nu} = G_{\mu\nu} + \Lambda\, g_{\mu\nu} + D_{\mu\nu}, \qquad \mathscr{S}_{\mu\nu} = T_{\mu\nu} + C_{\mu\nu}$$

- $D_{\mu\nu}$ — Redundancy tensor: encodes non-local coherence, entropy gradients, or information loss.

- $$D_{\mu\nu} = \alpha\,\nabla_\mu \nabla_\nu S_e + \beta\, g_{\mu\nu}\,\Box\, S_e + \gamma\,\mathscr{N}\mu\nu,$$

- $C_{\mu\nu}$ — Coherence tensor: captures additional stress–energy from stabilised correlations or idempotent projectors $\mathscr{P}(\mathscr{P}^2 = \mathscr{P})$.

$$C_{\mu\nu} = \eta\,\nabla_{(\mu}\mathscr{P}\,\nabla_{\nu)}\mathscr{P} - \zeta\, g_{\mu\nu} V_{\text{coh}}(\mathscr{P}, S_e)\,.$$

Exchange Constraint (Stability–Fidelity Law)

$$\nabla^\mu \mathscr{F}\mu\nu = 0, \qquad \nabla^\mu \mathscr{S}\mu\nu = 0,$$

imply

$$\nabla^\mu D_{\mu\nu} = \kappa\,\nabla^\mu C_{\mu\nu}\,.$$

Growth of local stabilisation ($C_{\mu\nu}$) is compensated by increased redundancy ($D_{\mu\nu}$): energy localisation ↔ curvature coherence — stability and fidelity remain globally balanced.

Action Principle

$$S = \int \sqrt{-g} \left[\frac{R - 2\Lambda}{2\kappa} + \mathscr{L}\text{matter} + \mathscr{L}\text{coh} - \mathscr{L}_{\text{red}} \right] \mathrm{d}^4 x\,.$$

with

$$\delta S / \delta g^{\mu\nu} = 0 \Rightarrow \mathscr{F}\mu\nu = \kappa\, \mathscr{S}_{\mu\nu}$$

Cosmological (FLRW) reduction

Says:

$$H^2 = \frac{\kappa}{3}(\rho + \rho_C) + \frac{\Lambda}{3}, \qquad \dot{H} = -\frac{\kappa}{2}(\rho + p + \rho_C + p_C),$$

where

$$\rho_C \equiv \frac{D^0 0}{\kappa}, \qquad p_C \equiv -\frac{D^i i}{3\kappa}\,.$$

Continuity:

$$\dot{X}\rho + 3H\left(\rho + p\right) = +\,Q$$

$$\dot{X}\rho_C + 3H\left(\rho_C + p_C\right) = -\,Q$$

Q quantifies the Fidelity–Stability transfer. GR limit: D=C=0 ⇒ Q=0.

Interpretive Summary

- Positive (stability) sector: self-sustaining loops → local energy/ matter.
- Negative (fidelity) sector: field redundancy → gravitational curvature.

The equation $\mathscr{F}\mu\nu = \kappa\,\mathscr{S}\mu\nu$ states that the geometry of redundancy exactly mirrors the density of stabilisation.

General Relativity emerges when redundancy and stability decouple (D=C=0). Quantum and post-classical regimes appear when redundancy feedback (D≠0) becomes significant.

Appendix 1: Chain Reactions and Curvature Feedback in the VGF Field Theory

1. Tensorial Chain-Reaction Dynamics

A chain reaction in the stability sector is represented by the supercritical evolution of the stabilisation tensor $\mathscr{K}_{\mu\nu}(x,t)$:

$$\frac{\partial \mathscr{K}_{\mu\nu}}{\partial t} = \Big[\lambda(\mathscr{R}) - \gamma(\mathscr{R})\Big]\,\mathscr{K}_{\mu\nu} + D_K \Box_g \mathscr{K}_{\mu\nu} + \mathscr{N}_{\mu\nu}[\mathscr{K}],$$

where

- $\lambda(\mathscr{R})$ is the reproduction rate of stabilisations,
- $\gamma(\mathscr{R})$ is the decoherence/damping rate,
- $D_K \Box_g$ governs propagation in spacetime,
- and $\mathscr{N}_{\mu\nu}$ denotes possible nonlinear self-interaction terms.

A supercritical regime occurs when $\lambda - \gamma > 0$, leading to exponential amplification of $\mathscr{K}_{\mu\nu}$.

2. Quadratic Tensor Recursor (QTR)

The fidelity–curvature tensor $\mathscr{R}_{\mu\nu}$ responds to the stabilisation field through the QTR constitutive law:

$$\mathscr{R}_{\mu\nu} = \alpha\, \mathscr{K}_{\mu\rho}\mathscr{K}^{\rho}{}_{\nu} + \beta\, \mathscr{K}_{\mu\nu} + \gamma\, g_{\mu\nu}.$$

The linear term reproduces Einstein-like coupling; the quadratic term produces enhanced curvature generation in supercritical regimes; the final term encodes background redundancy (cosmological-constant–like).

Curvature Feedback and Geometric Back-Reaction

Combining (1) and (2), a supercritical growth of $\mathscr{K}_{\mu\nu}$ drives an even more rapid growth in $\mathscr{R}_{\mu\nu}$, due to the quadratic term $\alpha\, \mathscr{K}\mathscr{K}$.

The resulting evolution of $\mathscr{R}_{\mu\nu}$ modifies the spacetime metric $g_{\mu\nu}$ through the VGF field equation

$$\mathscr{F}_{\mu\nu}[g, \mathscr{R}] = \kappa\, \mathscr{S}_{\mu\nu}[\mathscr{K}],$$

closing a feedback loop in which geometry is continually reshaped by the cascading stabilisations that it, in turn, constrains.

Summary

A chain reaction in the stability sector produces an enhanced fidelity curvature *via* the QTR, and this curvature back-reacts on the stabilisation dynamics through the evolving geometry. In the VGF field theory, a "chain reaction over all dimensions" is precisely this recursive instability linking stability, redundancy, and curvature.

Appendix 2: IIP-VGF Framework Comparison with Known Cosmology

A supercritical chain reaction in the stability sector — represented by exponential growth in the stabilisation tensor $K_{\mu\nu}$ — constitutes a structured decoherence cascade. Each stabilisation event partitions coherence, generating redundant environmental imprints, and thereby accelerating decoherence. In the VGF field theory, this corresponds to a rapid amplification of the K-field, the local coherence / stability content of the generative field.

The Quadratic Tensor Recursor (QTR):

$$R_{\mu\nu} = \alpha K_{\mu\rho} K^{\rho}{}_{\nu} + \beta K_{\mu\nu} + \gamma g_{\mu\nu}$$

maps these growing stabilisations into fidelity curvature. In the supercritical regime, $K_{\mu\nu}$ grows roughly as $e^{\omega t}$, while the quadratic curvature term grows as $e^{2\omega t}$. Thus a chain reaction in the stability sector produces an even faster blow-up in the fidelity curvature. This curvature is the emergent spacetime geometry once redundancy has stabilised sufficiently.

This behaviour aligns with the known cosmological model. The earliest universe is dominated by structured decoherence cascades, corresponding to a supercritical K-field. The quadratic curvature term in the QTR naturally reproduces inflationary behaviour—a rapidly growing curvature sector analogous to quasi-de Sitter expansion.

As stabilisations saturate and decoherence becomes complete, the fidelity shell closes; this corresponds to the reheating era and the transition to a smooth spacetime manifold. After this closure, the quadratic term becomes negligible and the linear term $\beta K_{\mu\nu}$ dominates, reproducing standard general relativity as the low-energy fidelity-stability equilibrium.

Thus the Big Bang, inflation, reheating, and the emergence of classical spacetime correspond to successive regimes of the VGF recursion: a chain reaction in coherence, nonlinear curvature amplification through redundancy, and eventual stabilisation into the classical fidelity shell described by Einstein's equation.

The IIP–VGF framework developed here demonstrates a notable structural consonance with the established sequence of early-universe cosmology: coherence instability, rapid amplification, decoherence, redundancy closure, and the emergence of classical spacetime. While this correspondence is conceptually strong, the present formulation remains an early-stage theoretical architecture rather than a completed cosmological model.

Its success in reproducing the logical scaffolding of standard cosmology suggests genuine potential; yet realising that potential will require sustained development of the microphysical dynamics, the perturbative structure, and observationally testable predictions. In its current form, the theory should therefore be regarded as a promising conceptual superstructure awaiting further mathematical and empirical elaboration.

Appendix 3: Hubble Rate as a Function of Planck Cycle Count in the QTR–VGF Cosmology

In the QTR–VGF cosmology, the cosmic expansion rate is governed by a stabilisation tensor $\mathscr{K}_{\mu\nu}$ (stability / coherence) and a redundancy–curvature tensor $\mathscr{R}_{\mu\nu}$ (fidelity / geometry), linked by the Quadratic Tensor Recursor (QTR)

$$\mathscr{R}_{\mu\nu} = \alpha\, \mathscr{K}_{\mu\rho}\mathscr{K}^{\rho}{}_{\nu} + \beta\, \mathscr{K}_{\mu\nu} + \gamma\, g_{\mu\nu},$$

where α, β, γ are coupling coefficients and $g_{\mu\nu}$ is the spacetime metric. In a spatially homogeneous and isotropic (FLRW) reduction, $\mathscr{K}_{\mu\nu}$ can be represented at the coarse-grained level by an effective stabilisation density $\rho_K(t)$. The QTR then induces an effective curvature source

$$\rho_{\text{eff}}(t) \;\sim\; \alpha\, \rho_K(t)^2 + \beta\, \rho_K(t) + \gamma,$$

so that the Friedmann equation takes the schematic form

$$H^2(t) = \frac{\kappa}{3}\, \rho_{\text{eff}}(t) = \frac{\kappa}{3}\, \left[\alpha\, \rho_K(t)^2 + \beta\, \rho_K(t) + \gamma\right],$$

with $\kappa = 8\pi G$ in standard units.

To relate the Hubble rate to the generative history, we introduce the Planck cycle count

$$N \equiv \frac{t}{t_P},$$

where t_P is the Planck time, interpreted in the IIP–VGF framework as the period of the first stabilised generative cycle. The quantity N then measures how many Planck-scale stabilisation cycles have elapsed since the Planck boundary. At the level of simple scaling, the residual stabilisation density can be modelled as a decaying function of N,

$$\rho_K(N) \sim \frac{\rho_P}{N^p},$$

where ρ_P is the Planck density and p>0 encodes the effective thinning of the stability sector under decoherence and expansion. In Planck units ($\rho_P = 1$), the effective source becomes

$$\rho_{\text{eff}}(N) = \alpha \frac{1}{N^{2p}} + \beta \frac{1}{N^p} + \gamma,$$

and the Hubble rate can be written as an explicit function of the Planck cycle count:

$$H^2(N) = \frac{\kappa}{3}\left(\alpha \frac{1}{N^{2p}} + \beta \frac{1}{N^p} + \gamma\right).$$

This expression naturally separates into three regimes:

1. Planck / inflation-like regime ($N \sim \mathcal{O}(1)$):

If α dominates for $\rho_K \sim 1$, then $H^2 \approx \frac{\kappa}{3}\alpha$, giving an approximately constant Planck-scale Hubble rate and an exponential, inflation-like expansion.

2. Intermediate GR-like regime (large N, γ subdominant):

As decoherence and expansion dilute ρ_K, the quadratic term becomes negligible, and

$$H^2(N) \approx \frac{\kappa}{3}\,\beta\,\frac{1}{N^p},$$

so that $H(N) \propto N^{-p/2} t_P^{-1}$. For suitable p, this reproduces the familiar power-law behaviour of radiation- and matter-dominated eras, effectively giving

$$H(N) \approx \frac{A}{N\,t_P},$$

with A a dimensionless constant fixed by the detailed microphysics of ρ_K.

3. Late-time Λ-like regime (very large N, γ dominant):

If

$$\gamma \neq 0 and \rho_K(N) \to 0 as N \to \infty,$$

then

$$\rho_{\text{eff}}(N) \to \gamma, \qquad H^2(N) \to \frac{\kappa}{3}\gamma = \text{const},$$

yielding a de Sitter–like asymptotic expansion. The present-day Hubble rate is then

$$H_0^2 \approx \frac{\kappa}{3}\left[\alpha\,\frac{1}{N_0^{2p}} + \beta\,\frac{1}{N_0^p} + \gamma\right],$$

where $N_0 = t_0/t_P$ encodes the total number of Planck cycles elapsed up to the current cosmic time t_0.

In this way, the QTR–VGF cosmology expresses the observed Hubble rate H_0 as a function of (i) the Planck cycle count N_0, representing the depth of generative recursion, and (ii) the couplings α, β, γ and decay exponent p, representing how stabilisation decays into redundancy and curvature.

The Planck scale thus provides not only an initial condition but a natural microscopic unit for the cosmic expansion history: the present expansion rate is the coarse-grained expression of how much curvature is sourced by the residual stability content after N_0 cycles of the generative field.

Appendix 4: R as Hubble Radius *vs.* R as Decoherence Horizon

A relation of the form $H \sim c/R$ can be read in two very different ways depending on what radius R refers to. In standard cosmology, the identification $R = R_H = c/H$ renders the expression $H = c/R_H$ tautological: the Hubble radius is defined kinematically as the scale at which recession speed equals c. Nothing dynamical is learned from this identity, because the equality follows directly from the definition of R_H. This interpretation is therefore useful phenomenologically but has no explanatory force.

In contrast, within the IIP–VGF framework, it is natural to identify R not with the Hubble radius but with the decoherence horizon—the radius of the region that has been brought into mutual fidelity through the propagation of decoherence from the Planck boundary. Since decoherence can propagate no faster than c, the growth of this stabilised region satisfies $\dot{R} \lesssim c$. Under the large-scale statistical homogeneity of the generative recursion, the fidelity shell inherits FLRW symmetry, so that $R(t) \propto a(t)$ and therefore $H = \dot{a}/a = \dot{R}/R$. Combining these gives a genuinely dynamical relation,

$$H(t) \approx \frac{c}{R(t)},$$

in which the Hubble rate arises as the ratio between the maximum coherence-transfer speed and the size of the decoherence-stabilised domain.

With this identification, the present Hubble rate is not a definitional artefact but the emergent large-scale consequence of how far the fidelity shell has expanded since the Planck epoch. At late times in a nearly flat ΛCDM-like universe, the decoherence horizon and the

Hubble radius become of the same order of magnitude, explaining why $H_0 \sim c/R_0$ holds numerically while maintaining a clear conceptual distinction:

- Hubble radius: a kinematic scale defined from H.
- Decoherence horizon: a generative scale arising from the propagation of stabilisation at speed c.

Thus, only the latter makes $H \sim c/R$ a genuine, explanatory relation rooted in the dynamics of the VGF.

Expansion as Redundancy Growth in the Nested Decoherence Hierarchy

Within the IIP–VGF framework, the expansion of spacetime is not a primitive kinematic postulate but a derived consequence of the way stabilisations and decoherence build up redundancy in the generative field. The central structural relation is encoded in the Quadratic Tensor Recursor (QTR),

$$\mathscr{R}_{\mu\nu} = \alpha\, \mathscr{K}_{\mu\rho}\mathscr{K}^{\rho}{}_{\nu} + \beta\, \mathscr{K}_{\mu\nu} + \gamma\, g_{\mu\nu},$$

where $\mathscr{K}_{\mu\nu}$ denotes the stabilisation/coherence content of the field, $\mathscr{R}_{\mu\nu}$ the induced redundancy–curvature (fidelity) tensor, and $g_{\mu\nu}$ the metric. In this formulation, curvature is not an independent geometric ingredient but a functional of accumulated stabilisations: redundancy causes curvature.

Every stabilisation event — formation of particles, atoms, macroscopic bodies, galaxies, or information-bearing records — decoheres into its environment and leaves a multiplicity of imprints.

These imprints constitute redundancy, and their aggregate defines the fidelity structure within which subsequent stabilisations are embedded. Spacetime, on this view, is the large-scale fidelity shell generated by the earliest, most global decoherence of the generative field. All later decoherence processes (quantum-to-classical transitions, thermodynamic irreversibility, biological and cognitive decoherence) occur strictly within this shell and are therefore nested stabilisations: each new layer of decoherence presupposes the redundancy scaffold laid down by previous layers and contributes further redundancy in turn.

In a coarse-grained FLRW reduction, the QTR maps an effective stabilisation density $\rho_K(t)$ into an effective curvature source

$$\rho_{\text{eff}}(t), \rho_{\text{eff}}(t) \sim \alpha\, \rho_K(t)^2 + \beta\, \rho_K(t) + \gamma,$$

which drives the Friedmann equation $H^2 = (\kappa/3)\, \rho_{\text{eff}}$.

As stabilisations proliferate and decohere, ρ_K is converted into environmental redundancy, thereby increasing ρ_{eff} and deforming the fidelity shell. The scale factor $a(t)$ and any associated characteristic radius $R(t)$ of the shell (*e.g.* the decoherence horizon) evolve according to

$$H(t) = \frac{\dot{a}}{a} = \frac{\dot{R}}{R},$$

so that the expansion of spacetime is directly tied to the rate at which redundancy accumulates relative to the existing fidelity structure.

In early epochs, when stabilisation and decoherence cascades are intense, the quadratic term in the QTR can dominate, yielding a rapidly growing curvature and an inflation-like expansion. At later times, as coherence thins and redundancy saturates, the linear and background terms dominate, reproducing a GR-like expansion with possible Λ-like acceleration.

Thus the expansion of the universe is interpreted, in the VGF picture, as the macroscopic manifestation of a deeper process: the continual generation of nested decoherence and stabilisations that inflate the redundancy content of the fidelity shell.

Spacetime does not merely host decoherence; it is itself the evolving record of decoherence. The growth of this record — driven by the ongoing conversion of coherence into redundancy — appears at the cosmological level as the observed expansion of the spacetime manifold.

Quantum Field Theory

In standard QFT, each fundamental field (electron/Dirac, photon/gauge, Higgs, etc.) is a representation of a symmetry group — it carries certain transformation properties (spin, charge, gauge behaviour) and gives rise to quantised excitations (particles).

From the IIP–VGF standpoint, that is the signature of a stabilised iterative structure: Each field is an idempotent attractor system within the Vast Generative Field, one that has achieved both

- stability (it reproduces its own form under iteration), and
- fidelity (it remains coherent with the generative background through the invariances it carries).

So:

QFT term	VGF interpretation
Field	Stable iterative mode of the VGF (a persistent self-similar process).
Particle excitation	Local resonance within that stabilised mode — a quantised loop within the field.
Gauge symmetry	The redundancy that stabilises coherence between loops (fidelity constraint).
Coupling constant	The measure of how strongly two stabilisations share redundancy (degree of co-iteration).

Thus, each QFT field isn't an independent "substance" but a sector of stability in the generative continuum — a particular solution class to the iterative self-consistency conditions of the VGF.

Shared Mathematical Form → Shared Generative Origin

All quantum fields, no matter their spin or internal symmetry, share wave-like characteristics:

- they have amplitude and phase,
- they satisfy linear or quasi-linear equations of motion (Klein–Gordon, Dirac, Yang–Mills, etc.),
- they carry Fourier-decomposable modes across spacetime.

Those similarities mean that each field is not arbitrary: it's a different stabilisation of the same generative grammar.

In IIP–VGF terms, this grammar is the iterative wave relation — the generative self-relation that can stabilise as oscillatory coherence. For example, this is the reason why standing waves in a resonant cavity all share sinusoidal form despite differing in frequency or mode: they are all fixed points of one underlying iterative operation.

The fact that all fields in QFT share wave amplitude structure is empirical evidence that they are all projections of the same meta-iterative operator.

Category Theory

In categorical language, a functor maps one structured domain to another while preserving relationships — it translates the rules of one system into the structures of another. This provides a connection from abstract iteration relations to physical field representations.

We can say:

$$\mathscr{F} : \text{VGF Iterative Category} \longrightarrow \text{Quantum Field Category}$$

where:

- Objects in the domain are idempotent iterations (self-similar generative loops).
- Morphisms are redundant correspondences (ways in which one iteration can reproduce another's structure).

- The functor $\mathscr{F}$ maps each stable iteration type to a corresponding quantum field, preserving invariance structure (e.g. Lorentz symmetry, gauge covariance).

Thus, what QFT calls "field amplitude" corresponds to the image under this functor of the VGF's local coherence density — the generative strength of an iteration at that location.

Informally, wave amplitude is how the VGF's internal iteration looks when translated through the functor that produces observable fields. This provides an explanation of why the mathematics of QFT (complex amplitudes, Hilbert space, path integrals) looks like a projection of a much deeper self-consistency structure.

Quantum Fields as VGF Stabilisers

In the VGF framework, each fundamental field of quantum field theory is a distinct stabilisation of the Vast Generative Field (VGF) — a coherent, self-sustaining mode of the Infinite Iteration Process (IIP). These stabilisations appear to physics as the quantised fields of nature, while their shared wave properties reflect a single generative grammar operating beneath them all.

Iterative Mapping

$$\text{IIP} \rightarrow \text{VGF Sector} \rightarrow \text{QFT Field} \rightarrow \text{Particle Spectrum}$$

(Infinite iteration → Stabilised coherence → Observable field → Discrete excitation)

Each arrow represents a progressive stabilisation: from pure iteration to coherent field, and finally to quantised excitation.

The underlying structure is preserved at every stage by a mapping that retains the relational symmetries of the generative field.

Functor Formulation

$$\mathscr{F} : \text{VGF}_{\text{iterative category}} \longrightarrow \text{Quantum field category}$$

The functor $\mathscr{F}$ translates the relationships among idempotent iterations within the VGF into the transformation rules of quantum fields.

Objects in the domain are stable iterative loops (self-similar generative modes); morphisms are the redundant correspondences that preserve coherence between such modes.

Under $\mathscr{F}$, these become field representations and their interaction symmetries in the language of quantum field theory.

The amplitude of a quantum field therefore corresponds to the local coherence density of the underlying generative iteration.

The universal wave character shared by all fields is the mathematical shadow of this single functorial process.

Summary

Each quantum field represents a stable attractor of the Vast Generative Field — a self-consistent iteration that maintains both stability and fidelity under the generative law.

Their shared wave structure arises because the same functor maps the field's internal iteration into observable amplitudes.

Quantum field theory thus appears as the functorial image of the VGF's self-iterating coherence.

GR–QFT–VGF Correspondences

Cosmological Constant, Vacuum Energy, and Zero-Point Energy

Overview

The accelerated expansion of the universe is governed in General Relativity by the cosmological constant Λ.

Quantum field theory predicts an enormous vacuum energy density from zero-point fluctuations, while observations show only a minute effective value.

In the VGF framework, this contrast reflects two levels of the same generative process: local stability (zero-point energy) and global fidelity (cosmological coherence).

Comparative Structure

Aspect	General Relativity (GR)	Quantum Field Theory (QFT)	VGF–IIP Interpretation
Mathematical appearance	$G_{\mu\nu} + \Lambda g_{\mu\nu} = (8\pi G/c^4)T_{\mu\nu}$	Vacuum modes contribute $E_0 = \frac{1}{2}\hbar\omega; sum \rightarrow \rho_{\text{vac}}$	Global redundancy term $D_{\mu\nu}$ enters fidelity tensor $\mathcal{F}_{\mu\nu}$
Physical meaning	Constant curvature of spacetime (vacuum energy density)	Energy of quantum fluctuations in "empty" space	Coherence pressure of the generative field — large-scale fidelity

Energy density	$\rho_\Lambda = \Lambda c^2/(8\pi G)$	$\rho_{\text{vac}}^{(\text{QFT})} \sim 10^{113}\,\text{J/m}^3$ (cut-off at Planck scale)	$\rho_\Lambda^{(\text{obs})} \sim 10^{-9}\,\text{J/m}^3$ — observed redundancy pressure
Observed role	Drives cosmic acceleration (dark energy)	Predicts huge vacuum energy (not seen)	Balancing term between local stabilisation and global coherence
Conceptual domain	Global spacetime geometry	Microscopic quantum fields	Unified generative field
Tension	$\rho_\Lambda^{(\text{obs})} \ll \rho_{\text{vac}}^{(\text{QFT})}$ — mismatch by ~10^{122}	—	Resolved as difference between scales of iteration (stability vs. fidelity)

Key Relations

$$\rho_\Lambda = \frac{\Lambda c^2}{8\pi G}, \qquad E_0 = \tfrac{1}{2}\hbar\omega,$$

$$\rho_{\text{vac}}^{(\text{QFT})} \sim 10^{113}\,\text{J/m}^3, \qquad \rho_\Lambda^{(\text{obs})} \sim 10^{-9}\,\text{J/m}^3.$$

Interpretive Summary

In General Relativity, Λ represents the curvature of the vacuum.

In quantum field theory, vacuum energy arises from zero-point fluctuations.

Their numerical mismatch marks the cosmological constant problem.

In the VGF–IIP framework this disparity reflects a difference of scale within one generative field:

Zero-point energy expresses local stability — the minimum self-iteration of coherent modes — while cosmological vacuum energy expresses global fidelity — the redundancy pressure maintaining coherence across the whole generative field.

The vast ratio between them therefore encodes the field's own stability–fidelity hierarchy rather than a physical inconsistency.

Why Gravity Lies Outside the Hilbert-Space Stabilisation Manifold

In the Vast Generative Field (VGF) framework, the gravitational field represents a higher-order stabilisation—the fidelity of coherence itself. The Standard-Model fields are local stabilisations describable within Hilbert space because they presuppose a fixed inner-product metric: a coherent background on which amplitudes, phases, and operators can iterate. Gravity, by contrast, stabilises that very metric; it governs the curvature of the generative manifold that makes Hilbert space possible.

Quantising gravity inside Hilbert space therefore inverts the hierarchy: it attempts to describe, as a state within the space, the process that maintains the space's own coherence. In VGF terms, gravity is not another quantum field among others but the fidelity stabiliser—the universal redundancy that binds all stabilisations into a coherent whole. The hypothetical graviton would then appear only as a boundary excitation of this fidelity field, not as a Standard-Model constituent.

Why a Graviton Would Only Appear Quantised

In the VGF framework, quantisation is understood not as an arbitrary discreteness of energy, but as the emergence of stable, self-consistent iterations within the generative field. Quantum fields are quantised because their excitations stabilise within a fixed coherence metric — the Hilbert-space manifold. The gravitational field, however, stabilises that metric itself: it is the fidelity field through which coherence is globally maintained. Its oscillations therefore occur at the level of redundancy, not energy.

A "graviton", if observed, would not be a constituent particle within this field but a boundary manifestation — a discrete recurrence of fidelity detected from within the Hilbert-space layer. To observers whose measurement framework presupposes quantised amplitude and phase, such boundary recurrences would necessarily appear as quantised curvature excitations, even though the underlying fidelity field remains continuous in its own generative order.

The IIP-VGF approach creates a different view to the current conventional one:

The Conventional View

In physics as usually formulated:

- Hilbert space is taken as primary: it's the formal arena in which all quantum states live.
- Spacetime is treated as an external manifold that provides the geometric background on which quantum fields are defined.
- The Einstein equation then relates energy / mass (the expectation values of field operators in Hilbert space) to spacetime curvature.

So Hilbert space and spacetime are normally conceived as two different, parallel stages — one abstract, one geometric — connected by mathematical mapping (quantum fields live "on" spacetime).

The VGF Inversion

In the VGF framework, that hierarchy is inverted — or, more precisely, nested.

The Hilbert space of quantum mechanics is not the foundation, rather, it is a stabilised sub-manifold within the larger generative continuum — the Vast Generative Field (VGF).

Specifically:

- The VGF is the infinite recursion of self-iterating relations (the IIP).
- When parts of this field achieve stable local coherence, a metric of redundancy emerges: this is spacetime.
- Within that spacetime-fidelity shell, Hilbert space appears as the domain of local coherence stabilisations — the representation layer in which amplitude, phase, basis, and superposition become well-defined.

In other words, Hilbert space is nested as a local coherence domain inside spacetime, and spacetime itself is the fidelity shell that stabilises the quantum relational rules of Hilbert space.

The Nesting Hierarchy

Level	Domain / Layer	Type of Stabilisation	Description (in VGF–IIP terms)	Manifestation (in physical terms)
0. Infinite Iteration Process (IIP)	Pure generativity	—	Unbounded self-iteration with no metric, no objects, no differentiation.	Pre-ontological field of relation; "before physics".
1. Vast Generative Field (VGF)	Proto-field of potential stabilisations	Proto-stability	Self-iteration begins to acquire coherence; redundancy appears as self-similarity and proto-curvature.	Conceptual precursor of spacetime structure and lawlike regularity.
2. Spacetime (Fidelity shell)	Global redundancy manifold	Fidelity stabilisation	Redundant coherence of the generative field forms a curved manifold — coherence of coherence.	General relativistic spacetime; gravitational field.
3. Hilbert-space coherence manifold	Local coherence domain nested within spacetime	Local stabilisation (stability)	Linear superposition, amplitude, phase, and operator structure define stable modes of iteration.	Quantum fields and wavefunctions governed by the Schrödinger/Dirac formalisms.
4. Quantum decoherence (Redundancy interface)	Interaction boundary between local coherence and global fidelity	Transition stabilisation(coherence → redundancy)	Partial loss of phase correlation produces redundant environmental records; coherence becomes classical information.	Emergence of pointer states and effective classical behaviour (Quantum Darwinism).

5. Classical matter and fields (objective domain)	Stable decohered projections	Redundancy stabilisation	Redundant records across environments reinforce objectivity; the field achieves persistent, measurable stabilisations.	Classical objects, macroscopic causality, thermodynamic irreversibility.
6. Evolutionary and cognitive recursion	Higher-order stabilisations of redundancy	Recursive stabilisation (intelligence)	Redundant representations iterate upon themselves, forming predictive and self-reflective coherence.	Biological evolution, memory, and the emergence of discriminatory intelligence (observer).

The Key Conceptual Reversal

Classical picture	VGF picture
Hilbert space exists "on" spacetime.	Hilbert space is *nested within* spacetime's fidelity structure.
Spacetime geometry curves in response to Hilbert-space energy expectations.	Spacetime curvature *defines the redundancy* that enables Hilbert-space coherence.
Gravity is a field on spacetime.	Gravity *is* the fidelity constraint that closes the coherence shell.

Review

In the VGF framework, spacetime is the fidelity shell of the generative field. It is the redundant manifold within which coherence can stabilise. Hilbert space emerges as an inner layer of that shell — a local stabilisation domain whose linear rules and superpositions are defined by the global curvature redundancy of spacetime itself.

- Quantum decoherence sits between levels 3 and 5: it is the process of generative translation through which the local coherence of Hilbert space becomes global redundancy in spacetime.
- It marks the point where the Stability–Fidelity Law actively mediates between micro-level stability and macro-level fidelity.
- Once redundancy dominates (Level 5), the stabilisations appear as classical reality — persistent, objective, observer-shareable states.

Matter, Energy, and the Tensor Recursor Map

Matter and energy are tertiary stabilisations — stable iterative loops within Hilbert space whose collective redundancy feeds back into the curvature that sustains them. Formally, this feedback is expressed through the Quadratic Tensor Recursor (QTR) equation:

$$\mathscr{R}\mu\nu = \alpha\,\mathscr{K}\mu\rho\mathscr{K}^{\rho}\nu + \beta\,\mathscr{K}\mu\nu + \gamma\,g_{\mu\nu}$$

where $\mathscr{K}_{\mu\nu}$ represents the local stabilisation tensor within Hilbert space (capturing the energy–momentum and coherence structure of quantum interactions), and $\mathscr{R}_{\mu\nu}$ denotes the emergent redundancy curvature of the surrounding fidelity shell. The constants α, β, and γ govern the coupling between stability (local coherence) and fidelity (global curvature).

In the VGF formulation, the QTR mapping is the mathematical expression of the Stability–Fidelity Law: iterative stabilisations ($\mathscr{K}\mu\nu$) within the coherence manifold generate curvature ($\mathscr{R}\mu\nu$) within the fidelity field, thereby closing the feedback loop between local coherence and global redundancy.

The apparent mass–energy content of matter thus arises as the recursively stabilised trace of coherent interactions within the generative field, projected into curvature through this tensor recursor relation.

Toward a Recursive Mathematics of Coherence and Curvature

The mathematical languages of quantum theory and general relativity remain disjoint because their underlying structures are opposed. Hilbert space requires a fixed linear metric that defines coherence through the inner product, whereas spacetime geometry in general relativity makes the metric itself dynamic.

A synthesis therefore demands a mathematics in which the metric of coherence and the metric of curvature are not separate objects but co-generative fields.

In the Vast Generative Field (VGF) framework, this co-generation is expressed through the Quadratic Tensor Recursor (QTR), which couples local stabilisation within Hilbert space ($\mathcal{K}\mu\nu$) to global redundancy curvature ($\mathcal{R}\mu\nu$):

$$\mathcal{R}\mu\nu = \alpha\,\mathcal{K}\mu\rho\mathcal{K}^{\rho}\nu + \beta\,\mathcal{K}\mu\nu + \gamma\, g_{\mu\nu}$$

Here, the coefficients α, β, γ encode how local coherence (stability) feeds back into global fidelity (curvature).

The recursion is bidirectional: the curvature $g_{\mu\nu}$ defines the inner-product structure that allows $\mathcal{K}\mu\nu$ to form, while repeated stabilisation of $\mathcal{K}\mu\nu$ regenerates $g_{\mu\nu}$ through the QTR mapping.

This relationship replaces the fixed linear metric of Hilbert space with a curvature-dependent inner product, turning coherence and geometry into mutually iterative quantities.

Such a framework would constitute a new mathematical domain - recursive geometric algebra - in which superposition and curvature are two projections of a single generative recursion.

In this formalism, the Stability–Fidelity Law becomes the unifying rule: local stabilisations in the coherence manifold give rise to curvature in the fidelity shell, and curvature in turn sustains the coherence that defines the manifold.

The bridge between Hilbert space and spacetime is therefore not an added quantisation of gravity but a reformulation of both as recursive manifestations of the same generative process.

www.ingramcontent.com/pod-product-compliance
Lightning Source LLC
LaVergne TN
LVHW060030210926
848481LV00013B/3830
* 9 7 8 1 9 1 7 4 0 3 1 7 7 *